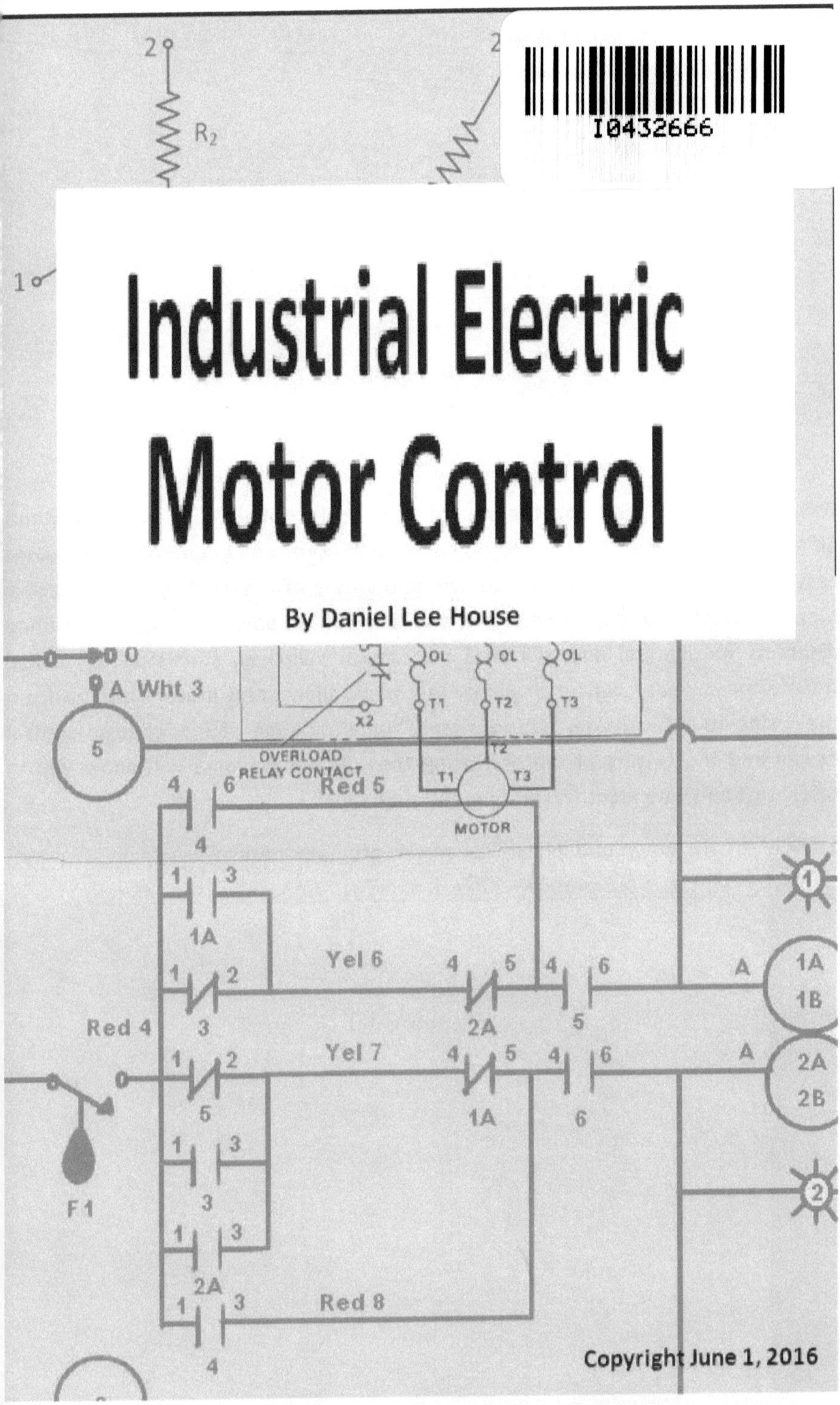

Industrial Electric Motor Control

By Daniel Lee House

Copyright June 1, 2016

This book is designed for the novice and professional alike, to understand different types of motors and switchgear used at home and in industry. It provides easy to follow diagrams for the control and wiring of different types of electric motors including Ac/Dc, single phase, and three phase power. Included are wiring diagrams for manual and electrical mechanical switches, start-stop, reversing, transformers, phase converter plans, test panel plans, and motor hook-up from the inside to the outside, with a simple understanding of ladder logic control design and the testing of motors. Using these building blocks will allow you to safely test and wire electrical equipment of all types.

2 HP – 30 HP download phase converter program here: https://kat.cr/phase-converter-windows-program-by-convertadyn-2015-house-t12517717.html

Industrial Electric
Motor Control

By Daniel Lee House

Copyright June 1, 2016

Industrial Electric Motor Control

This book is dedicated to those that need to know.

Series Circuits

To fully understand the uses and dangers of electrical power one must be familiar with some basic math. To start with we will look at a series circuit. Remember that in series circuits the amps are constant. The sum (adding) of the voltage drops in a series circuit is equal to the input voltage.

Look at the diagram below. Notice that each resistance (load) is in series

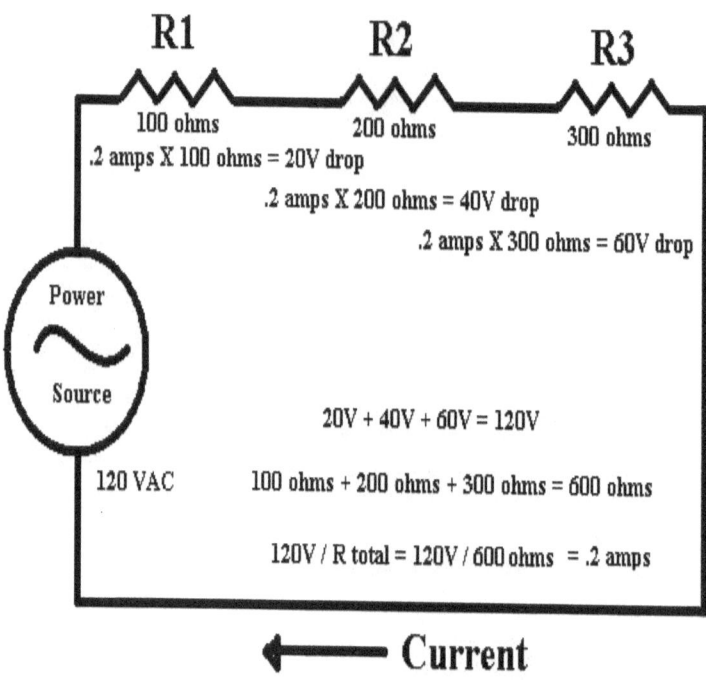

connected end to end.

Now look at the math:

$V/R = I$

V = Voltage   R= Resistance (ohms)   I = Current

Knowing only the resistance of one load in a series circuit and the voltage drop across that one load allows you to know the current in the series circuit ($V / R = I$).

Knowing the voltage drop across the one load and the current allows you to know the one load resistance ($V / I = R$).

Knowing the one load current and the one load resistance allows you to know the voltage drop across the one load ($I \times R = V$).

Resistance in series is added together R1 + R2 + R3 + Rn..........= R total

R1 = 100 ohms

R2 = 200 ohms

R3 = 300 ohms

= 600 ohms

This gives us a constant formula for understanding the voltages you are working with and their requirements. This is Ohm's Law known as the voltage triangle

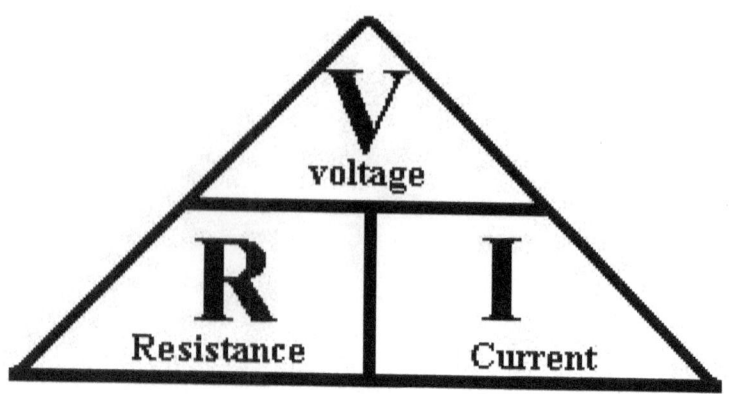

Parallel Circuits

The next circuit is a parallel circuit. Notice that the voltage across each load is always equal to the input voltage. The current can be different in parallel circuits if the resistance is different. The sum (added) load currents is the total load current $I1 + I2 + I3 + In... = I$ total

1.2 Amps

.6 Amps

<u>.4 Amps</u>

2.2 Amps

Ohm's Law is very useful and never changes.

Voltage is like pressure. Think of a water hose. A smaller water hose (smaller wire) supplies less water (current) to an area (load) and a larger water hose supplies more water (current) to an area (load). When wire is used in service it becomes a conductor. The wire carries the load current and the insulation protects from short circuit breakdowns. It is important to have large enough wire or conductor and the proper voltage rating, of conductors, to avoid fires or meltdowns.

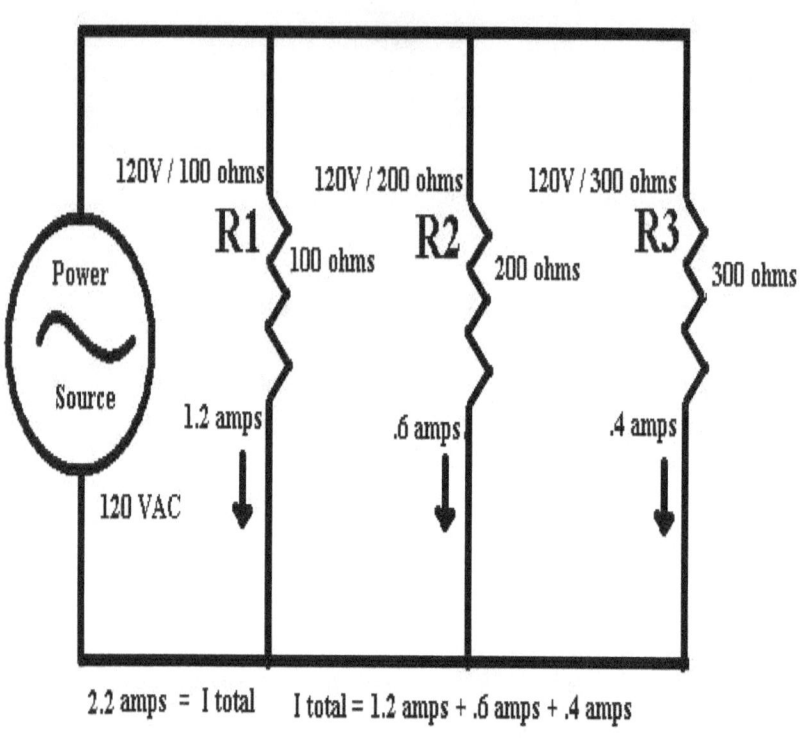

Wire Size and Current

Remember Ohm's Law and these basic wire sizes for your work. And notice that it always takes two conductors to energize electrical equipment of all types, that means there must be at least two ends (the start and the finish of the conductor).

Wire size to fuse size

#18 AWG wire – 5 amps

#16 AWG wire – 10 amps

#14 AWG wire – 15 amps

#12 AWG wire – 20 amps

#10 AWG wire – 30 amps

#8 AWG wire – 40 amps

#6 AWG wire – 60 amps

Single Phase AC Power

Think of alternating current (AC) as a voltage that goes from a low negative level to a high positive level at 60 times a second. This produces a buzzing and vibrating effect on AC equipment that can be heard and felt.

Home wiring (single phase) is fairly easy but one should follow applications used in the National Electric Code (NEC) for safety. Single phase power consists of 2 (hot) leads, 1 neutral lead, and 1 ground lead (the neutral and the ground should be connected or bonded together in the main power panel). The 2 (hot 240 volt) leads are usually red and black, the neutral lead is white always, and the ground lead is either bare un-insulated wire or green always.

240 volts is connected between to the red and black.

120 volts is connected between the red (hot) and white (neutral) or the black (hot) and white (neutral). Usually only black and white are used for 120 volt. If white is used for a hot lead at any time it must be marked with black tape.

The ground lead is connected to the neutral bar in the load center and also connected to an outside grounding rod through the ground.

There are many types of connectors for 240 volt applications but usually only one type for 120 volt applications.

120 volt applications use a polarized (large spade silver colored), a hot (small spade brass colored), and a ground (half moon slot green colored). The wires connected to plugs and outlets must always be correct.

240 volt applications use 2 hot leads (red and black brass colored) and a ground lead (green colored) for standard applications.

Both 120 volt and 240 volt systems use special wiring sometimes, depending on load requirements, that use a 4-wire system. A 4-wire system has 2 hot leads – 1 neutral lead – 1 ground lead.

It is important to be able to test all types of electrical equipment safely so you will need to build a circuit test panel like the one that follows. This allows you fast safe connections to your work with controlled currents to prevent short circuits.

This test panel allows testing for grounds, short circuits, opens, or normal operation safely. Each button on the panel performs a function of the test panel. Each time a button is selected a panel light is illuminated beside the button. First is the OFF button to turn off all functions, next the LIGHT Test Button (with neon lamp), next a LOW Test button (10 amp limited current for small equipment), next a HIGH Test button (20 amp limited current for larger equipment), next a Full voltage 120 V volt button, and a Full voltage 240 V button. All of these buttons are connected and controlled to a single 3-wire test cable with alligator clip ends. All five Modes can be turned off with the use of the OFF Mode button or any Mode may be selected randomly and the previous Mode will be automatically turned off.

Test Panel Components List

0, 1, 2, 3, 4, 5, 6, 7, 8, 9, 10, 11 – control relays

24 volt ac coil

5 amp contacts

3PDT (3 pole double throw)

12 – control relay

24 volt ac coil

5 amp contacts

1PST (1 pole single throw)

6 – momentary push buttons

5 amp

6 – panel indicator lamps 24 volt

1 – neon panel lamp 120 volt

2 – 1200 w duct heater elements (higher heat) 120 volt or 2400 w elements (lower heat) 240 volt

T1 – 40w 120v / 24v transformer

T2 – 10w 120v / 24v transformer

F1 – 5 amp 250v

13, 14 – control relays

120 volt coil

10 amp contacts

3PDT (3 pole double throw)

M1 – contactor

120 volt coil

30 amp contacts

2PST (2 pole single throw)

1 N.C. (normally closed) auxiliary contact.

M2 – contactor

120 volt coil

50 amp contacts

2PST (2 pole single throw)

1 N.C. (normally closed) auxiliary contact.

1 – 50 Amp AC Panel Meter

1 – 60 amp 240 volt service

#6awg input wire for 240 volt

#10awg output 3-wire rubber covered cord

3 – Alligator Clips

1 – 14 X 20 X 4 enclosure

The function of each circuit is as follows:

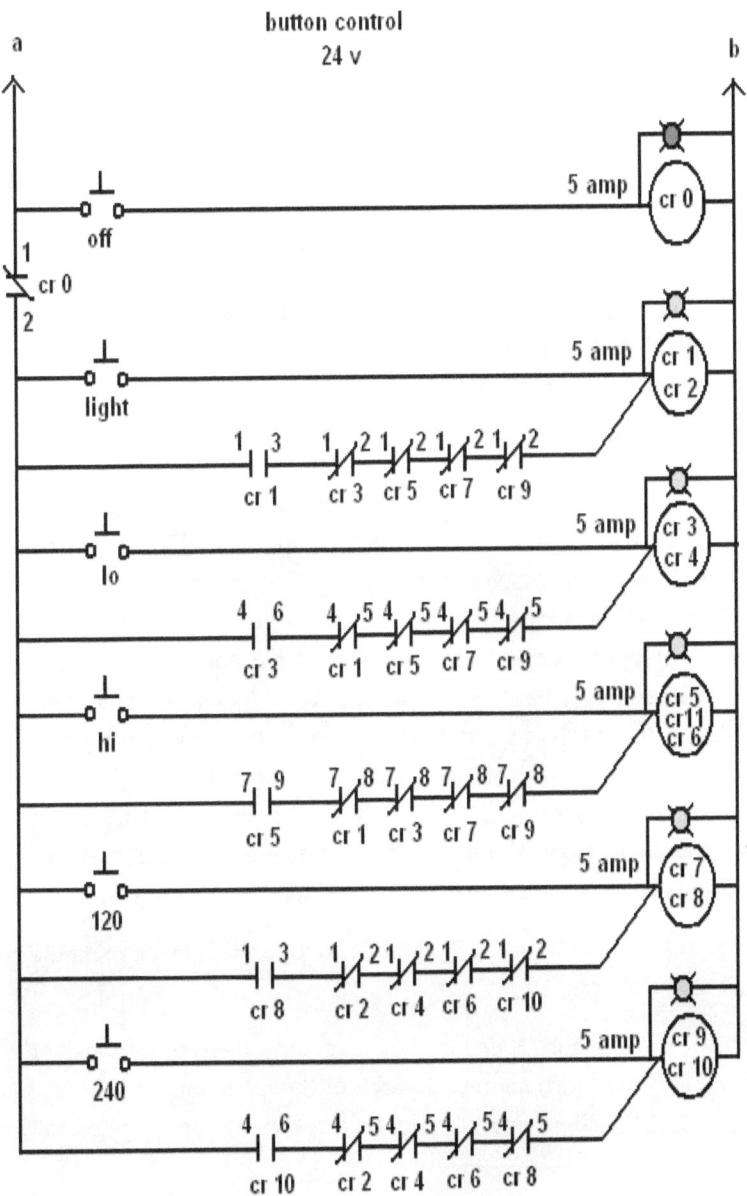

Diagram 1

TEST1 is the main control circuit. Voltage is applied to L1 and L2, I used 24vac relays for the control relay coils but you can easily modify for 120vac or 240vac coil relays. The six momentary panel buttons begin in each control mode. The top button is for OFF (stop), next is TEST LIGHT, then, LO TEST, HI TEST, 120 VOLTS, and finally 240 VOLTS. The OFF momentary button is connected to a single control relay '0'. In the diagram '0' is the relay number, 'L' is a panel light (same voltage as the coil), C1 and C2 are the coil number leads on the relay I used. L1 is connected to a normally closed contact (1 and 2, according to my relay type.) on the '0' relay which performs shut down of all modes. The next 5 push buttons each have 2 relays connected together to make a 6 pole relay; for example the next relay marked LIGHT has relays 1 and 2 with 'L' (panel light), LO has relays 3 and 4 with 'L', HI has relays 5 and 6 with 'L' and '11' (an extra relay added to be used to turn on LO test with HI test mode in TEST2 diagram), 120V has relays 7 and 8 with 'L', 240V has relays 9 and 10 with 'L'. At the top of each contact in the diagram is the pin number of the relay contact and under the contact is the relay number. There are a lot of connections but the operation is fairly simple.

TP3 (test point) - Relay '0' turns off all modes of operation by opening the current path to all relays '1' thru '10'.

TP4 turns on relays '1' and '2' and latches to relay '1' through terminals '1' and '3' and a normally closed (off) contact on each of the other modes terminals '1' and '2' (relays 3, 5, 7, 9). When relays '1' and '2' closes it also opens a normally closed contact connected through each of the other modes (relay '1' terminals '4' and '5' in LO mode, relay '1' terminals '7' and '8' in HI mode, relay '2' terminals '1' and '2' in 120V mode, and relay '2' terminals '7' and '8' in 240V mode) which un-latches all other modes. Each mode is connected in the same manner to allow switching directly to each mode from another mode without switching OFF first.

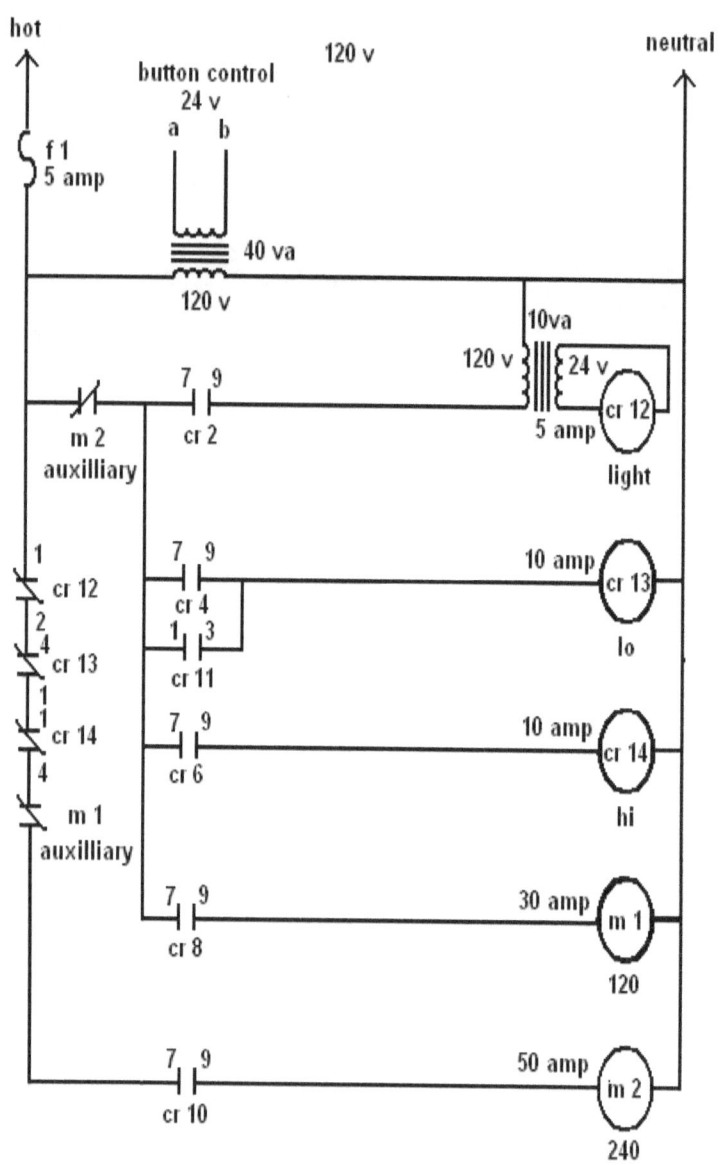

test 2 Daniel Lee House Copyright 2016

Diagram 2

TEST2 is the main line switching relay and contactor diagram for the coil and auxiliary circuits. It would be easier to modify the diagram and use all the same voltage coils for each relay and contactor but I used different voltages based on the parts as an example of multiple power supplies. The 24 volt power supply, at the top, supplies power to the TEST1 diagram. Relay '12' controls the test LIGHT, '13' is a relay to control the LO test, '14' is a relay to control the HI test, 'M1' is a contactor to control the 125V output, and 'M2' is a contactor to control the 240V output. These relays and contactors are interlocked (meaning operation depends on the output of another operation) to prevent 120V outputs from crossing the 240V output. If any of the 120 volt modes are in use, the current path is interrupted to the 240 volt output 'M2' through relay '12' terminals '1' and '2', relay '13' terminals '1' and '4', relay '14' terminals '1' and '4', and contactor 'M1' normally closed auxiliary contact. If contactor 'M2' (240V mode) is turned on, the current path is interrupted to all of the 120 volt modes through a normally closed auxiliary contact on 'M2'.

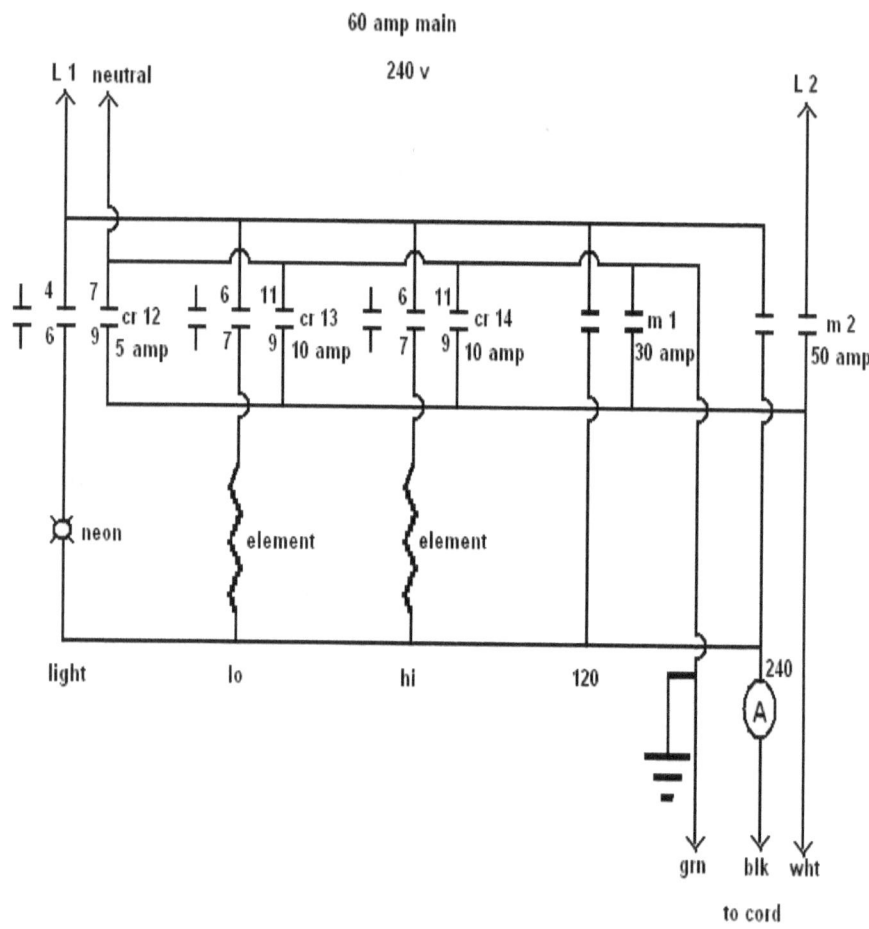

Diagram 3

TEST3 is the line connection diagram for input and output voltages to the test panel. The input is Black, Red (hot 240v), White (neutral) connected to grounded panel. Mode 2 (LIGHT) relay '12' is placed in series through the output test leads, with a neon light on the panel to check for moisture or current. A neon lamp is sensitive to very low currents and will light dim to show good insulation and brighter for bad insulation or moisture. Mode 3 (LOW) relay '13' is placed in series through the output leads, with one element for 10 amp limited output. Mode 4 (HIGH) relay '14' is placed in series through the output leads, with one element for 10 amp maximum output which is added to the '13' relay for 20 amp maximum output ('M2' and '14' operate together in this mode). Mode 5 (120 V) contactor 'M1' supplies full 120 volt output to the test leads. Mode (240 V) contactor 'M2' supplies full 240 volt output to the test leads. The load flows through a 50 amp meter, in series with the black output lead on the panel, for reference to current draw in any mode. The output is a 3-wire #10 AWG rubber covered cord with alligator clips at the ends.

The two current heater elements are mounted away from the test panel.

11 PIN   **.187" TAB**

```
    4 5 8  -  N.C.  -  8 5 2
    3 7 9  -  N.O.  -  9 6 3
    1 6 11 -  COM   -  7 4 1
    2  10  -  COIL  -  C2 C1
```

Test4

DECEMBER 1995 DLH

Operation

This test panel allows testing for grounds, short circuits, opens, or normal operation safely. Each button on the panel performs a function of the test panel. Each time a button is selected a panel light is illuminated beside the button. First is the OFF button to turn off all functions, next the LIGHT Test Button (with neon lamp), next a LOW Test button (10 amp limited current for small equipment), next a HIGH Test button (20 amp limited current for larger equipment), next a Full voltage 120 V volt button, and a Full voltage 240 V button. All of these buttons are connected and controlled to a single 3-wire test cable with alligator clip ends. Input voltage is 4-wire 240 volt single phase. Output is 2-wire with ground. At the top of the panel is an Ampere meter for actual current readings for the output. All five Modes can be turned off with the use of the OFF Mode button or any Mode may be selected randomly and the previous Mode will be automatically turned off.

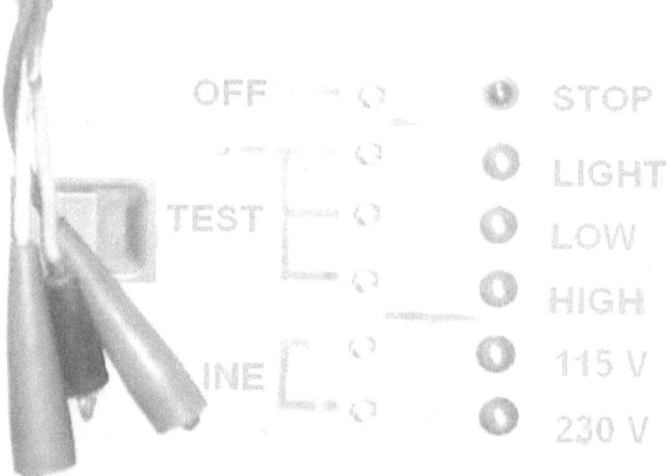

Alternating Current AC Electric Motors

Starting with electric motors there are several different types in use. When the rotating part has a rotating winding and commutator, it is called an armature AC or DC, the inside rotating part on other motors are called rotors AC only. The stationary windings are in the stator. Motors come in all types of enclosures both open and enclosed. There are single phase [2-wire] and three phase [3-wire] power AC motors. Some single phase motors have internal or external parts like, thermal protection, switches, capacitors, and relays. Capacitors can be tested by removing and applying a low current charge across the terminals for 1 – 2 seconds, then remove charge for 5 seconds, and short terminals with a wire or screwdriver. The capacitor should hold a strong charge and should not be bulged, cracked, burnt-up, or exploded.

Some motors are ball bearing and others are oil bushing bearing types. Checking for wear is the first check. Check the motor shaft that is spins freely without end play for ball bearings or up and down play and from side to side across the shaft. If there is noticeable movement then it is a problem.

Next you are ready to begin testing the electrical parts. It is important to know what type you are working on. Choose the type of motor to learn about internal parts and wiring. Carefully inspect windings for corrosion, cracked insulation, burns [carbon], and dirt.

Single Phase Electric Motors

Skeleton or Shaded Pole motors

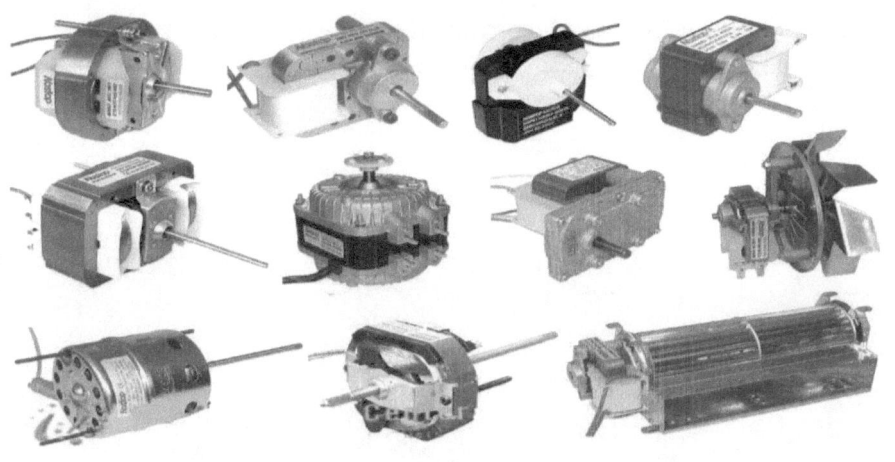

This type of motor is used on small fans, humidifiers, ice makers, blowers, etc. for low torque starting fixed rotation. Because it is usually very open to the windings gives it its name skeleton. To change rotation you must dismantle the motor and reverse the rotor shaft and possibly end housings also.

There is only one power winding and rotor with no internal parts. Usually one or two speed motors like the diagram below. The Lo speed winding adds resistance to the motor and reduces power and speed.

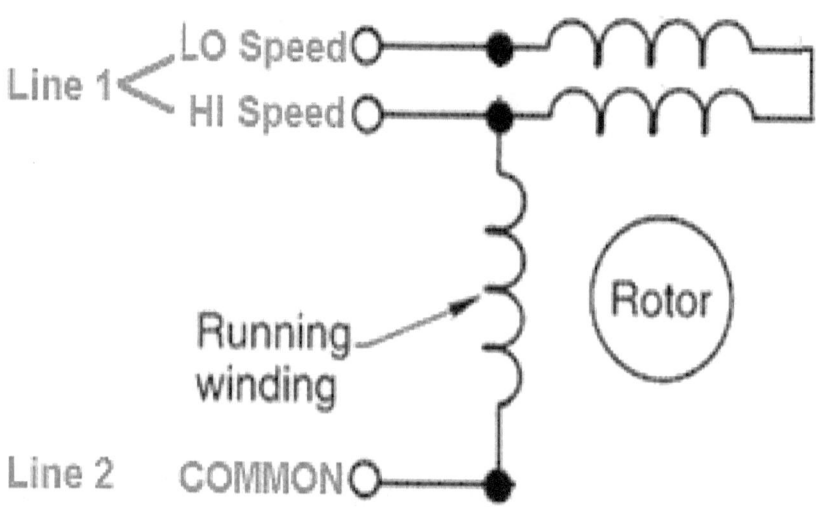

Permanent split capacitor

This motor uses a running capacitor for low torque starting reversible rotation. The capacitor is usually externally mounted. There are two separate windings, one for start and one for run. Notice the configuration in the diagram below. Here the start (auxiliary) winding is connected across the running (main) winding in series through a low current (run) capacitor. This adds to the power of the motor. This motor can be reversed by interchanging the running winding leads.

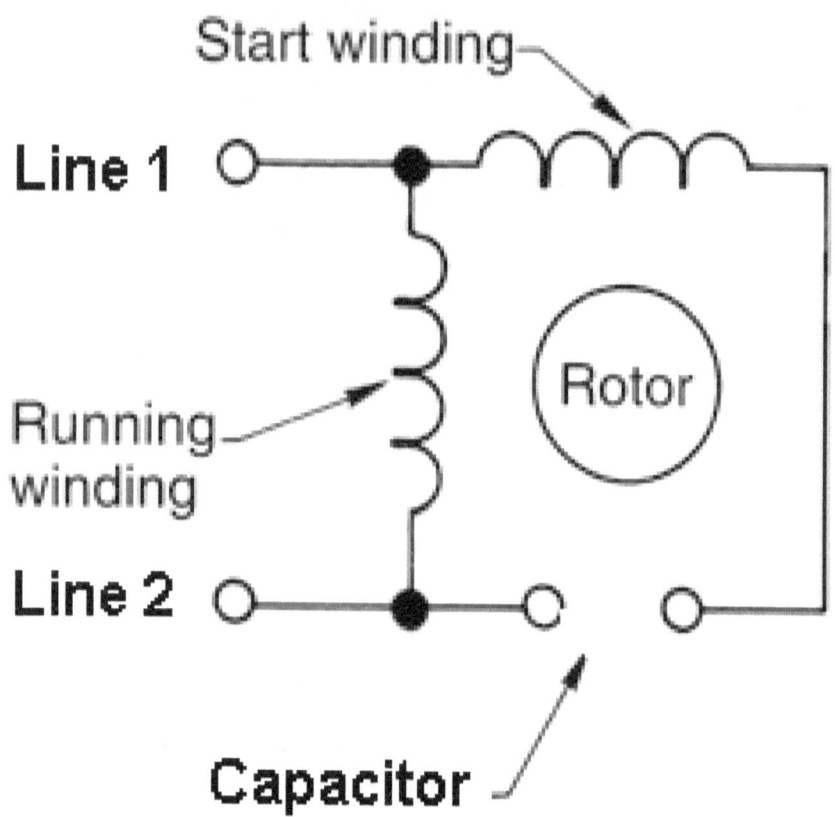

Split phase

This motor has an internal starting switch for moderate torque starting reversible rotation. There are two separate windings, one for start and one for run. The diagram below shows the configuration. When the motor starts the centrifugal switch is closed and causes a high current to the motor. When the motor reaches about 75% of speed, the switch opens and the motor continues to operate on the running winding only. This motor can be reversed by interchanging the running winding leads.

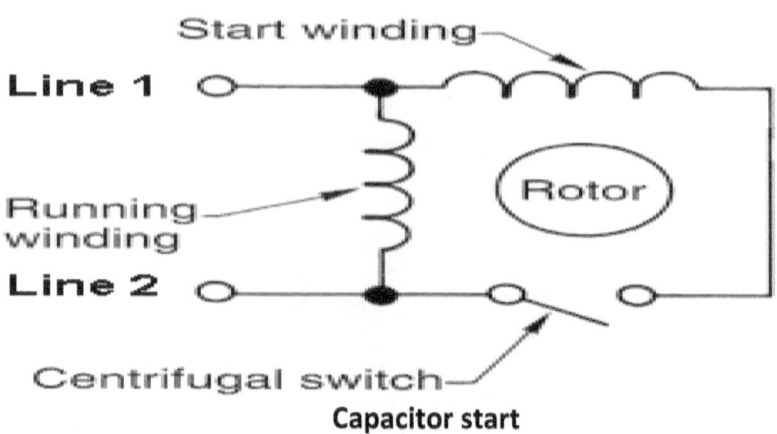

Capacitor start

This motor usually has an external starting capacitor can for high torque starting reversible rotation. The configuration in the diagram below shows a split phase motor with a high current start capacitor in series with the start winding and the centrifugal switch. This causes the motor to produce higher torque. Like the split phase motor when the centrifugal switch opens the motor operates only on the running winding. This motor can be reversed by interchanging the running winding leads.

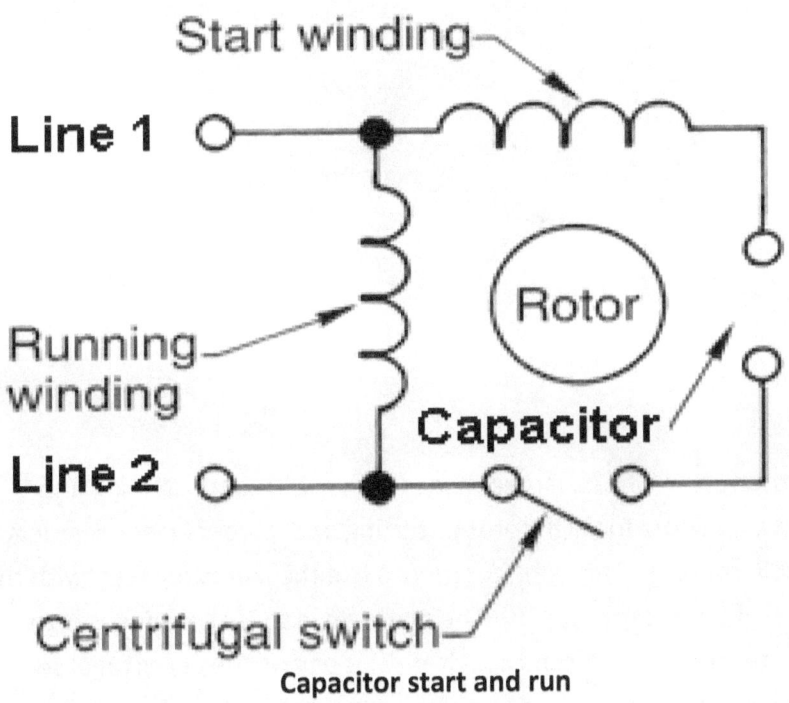

Capacitor start and run

This motor usually has a larger capacitor can with both starting and running capacitors for high torque starting and lower current running reversible rotation. This configuration is like the capacitor start with the addition of a run capacitor (B). The start capacitor (A) produces high starting torque and the run capacitor (B) is across the centrifugal switch and allows the start winding to aid the running winding. This motor can be reversed by interchanging the running winding leads.

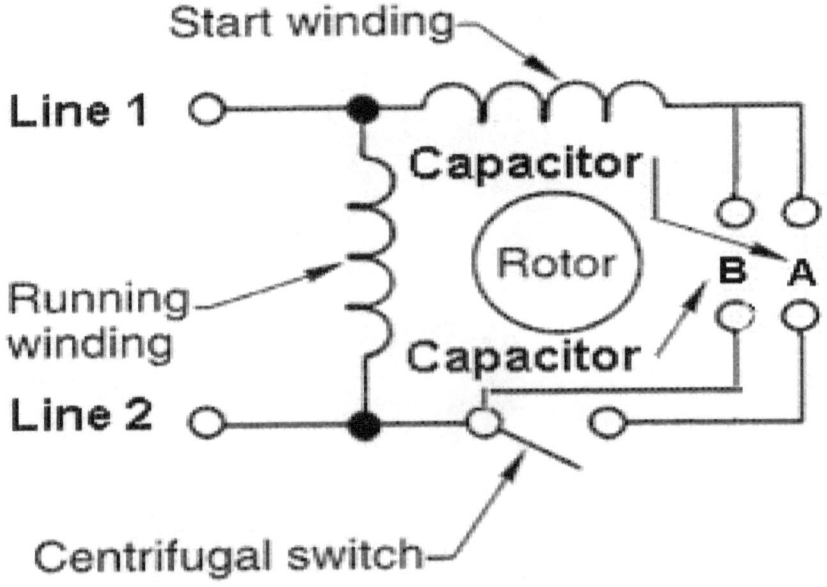

Series Ac volts or DC volts

This motor has a power winding with an armature and commutator brushes. Used on drills, vacuums, etc. This motor can be used on AC or DC power at the rated voltage with reversible rotation. Be careful, not all brush motors are AC and DC. The stator winding must be in series with the armature for this type. The wire size on the stator winding is about the same size as the armature winding size because each carries the same amount of current. A set of carbon brushes connects to the brush holders to make contact with the armature. See the configuration diagram below. This motor can be reversed by interchanging the stator leads.

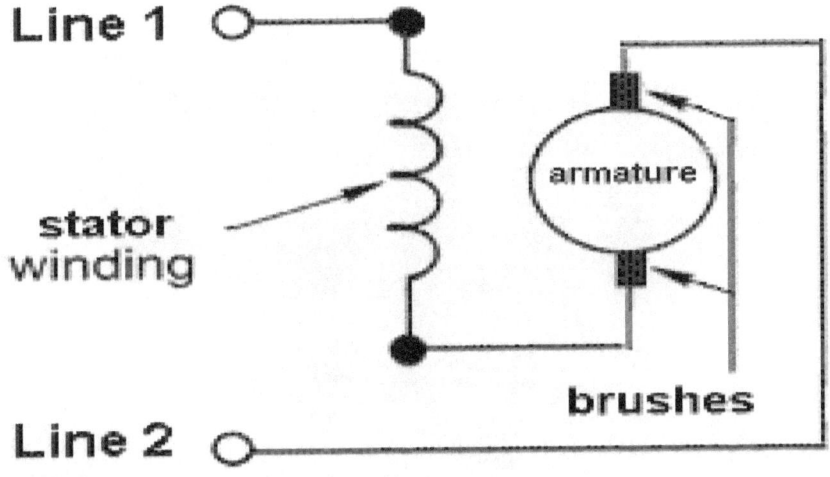

Three Phase Electric Motors

These are industry motors that require three AC power leads to operate and come in all sizes and voltages with reversible rotation. This motor has a rotor and a stator winding and uses less power than single phase motors. There are two types of connections for this motor, one is called Wye or Star and the other is called Delta. The resistance between windings should be equal. See the configuration Diagrams below. To reverse rotation for either type interchange any two line leads R, S, T.

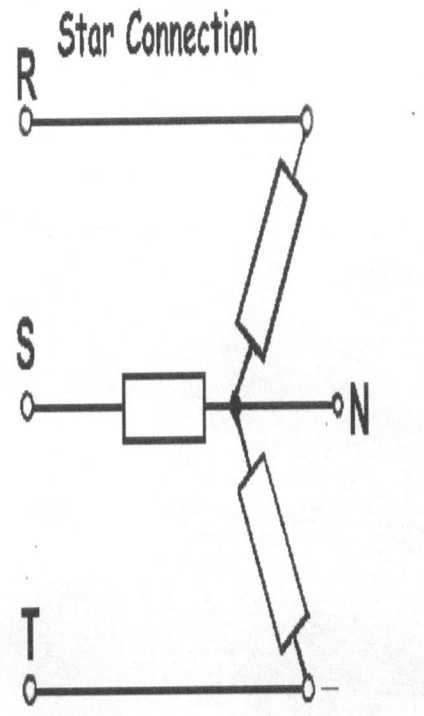 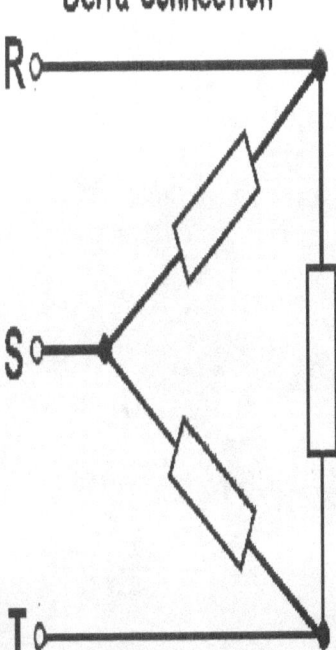

Direct Current DC Electric Motors

These are motors for use on DC power only and come in all sizes and voltages with reversible rotation. Direct Current (DC) motors have a positive (+) line and a negative (-) line voltage. They may have internal windings called fields or ceramic magnets and an armature winding. To reverse rotation interchange the armature leads. Motors with ceramic magnets only have two leads (+) and (-). Because DC is direct current you usually will not notice any buzzing or vibrating noise associated with operation.

Transformers

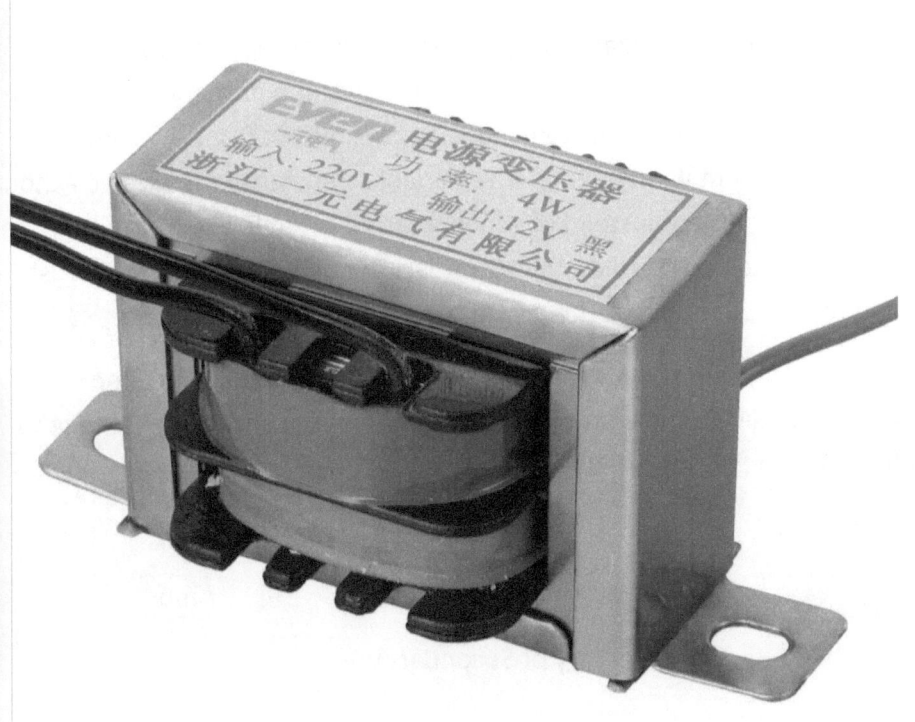

Transformers are used to increase or reduce output voltages for voltage control. There are both single phase and three phase transformers made in three different types. The diagram below shows the configuration for the three main types. Transformer theory controls the output voltage where Turns Input (N1) / Voltage Input (V1) is equal to Turns Output (N2) / Voltage Output (V2).

$$\frac{N1}{V1} = \frac{N2}{V2}$$

The Isolated transformer has a primary winding that is not connected to one or more secondary windings which may or may not be connected together. The Auto transformer has the primary and secondary winding connected and voltage is output from the common lead to different winding taps. The Variable transformer is like the Auto transformer except instead of winding taps a variable wiper arm is connected to each turn on the secondary winding using a carbon brush and the output is across the common lead and the wiper brush.

Transformers are usually rated in watts only so to determine amp rating you divide the watt rating by the primary or secondary voltage to find the amps.

$$\frac{\text{Watts}}{\text{Volts (Primary or Secondary)}} = \text{Amps}$$

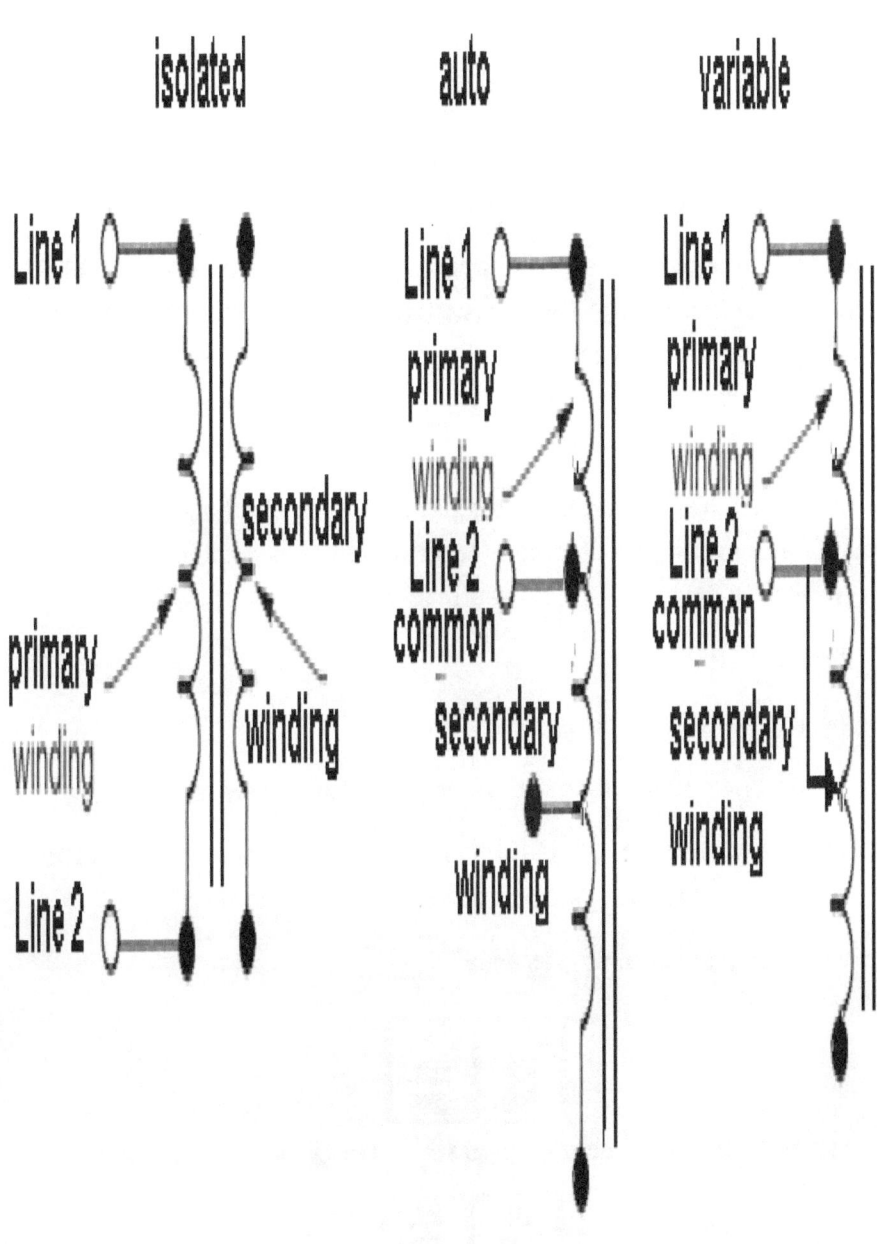

AC and DC power use many different types of electric controlled switches or solenoids to control electric power and motion. All of these types whether single phase or three phase or DC use only two coil winding ends to provide a current path for operation. The important thing to remember is to use proper voltage and current ratings for each device used.

Switches

Switches come in a variety of shapes and sizes, but the operation is the same depending on the type of switch. There are manual/mechanical and electrical/mechanical types of switches. Manual switches, by name, use any form of manual activation while electrical switches are controlled by electric activation.

Manual switches are types of toggle, rotary, or slide activation. Terms used are the number of poles (current paths), available positions for operation (throw), and special features. For example:

Single Pole (SP)

Single Throw (ST) positions

Or (SPST) single pole single throw, or just an on-off switch.

This is how switches get their names according to the National Electrical Manufacturers Association (NEMA) Standards.

Common toggle switch types are:

DPST double pole single throw

DPDT double pole double throw

SPST single pole single throw

SPDT single pole double throw

3PST three pole single throw

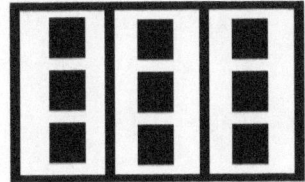

3PDT three pole double throw

SPDT C/O - single pole double throw center off

DPDT C/O - double pole double throw center off

3PDT C/O - 3 pole double throw center off

Typical wiring of these types of switches are:

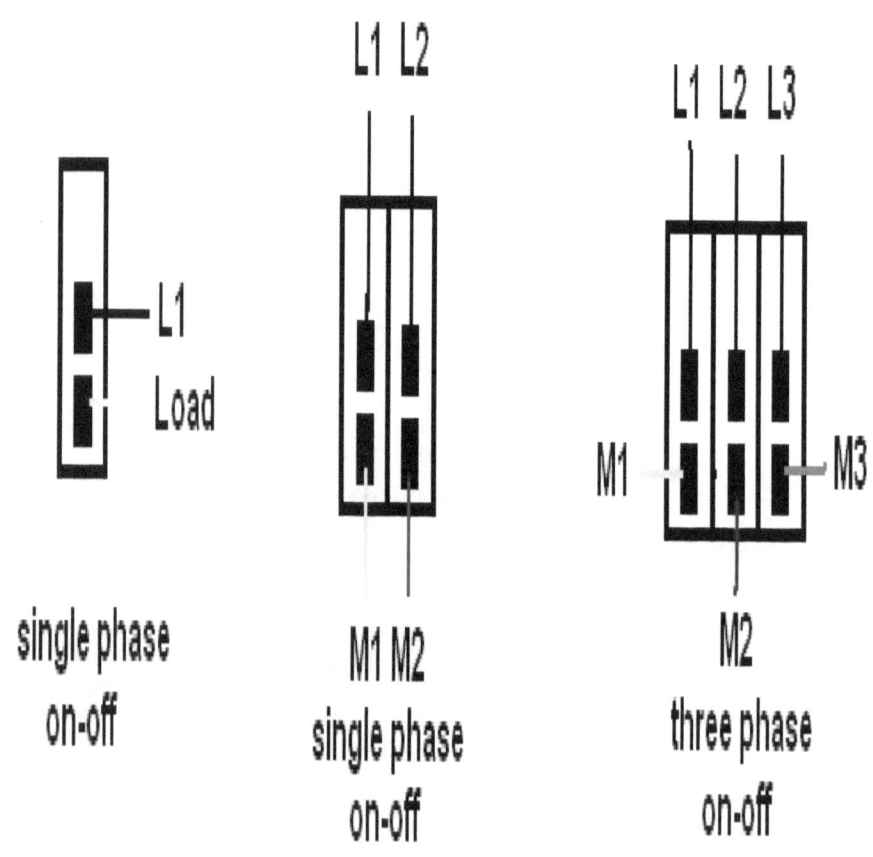

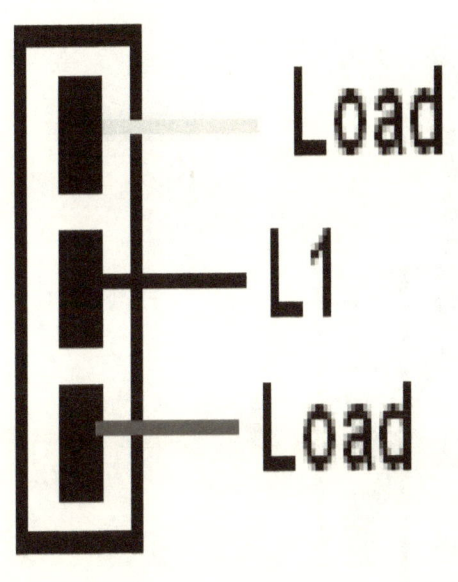

on-on

center-off   on-off-on

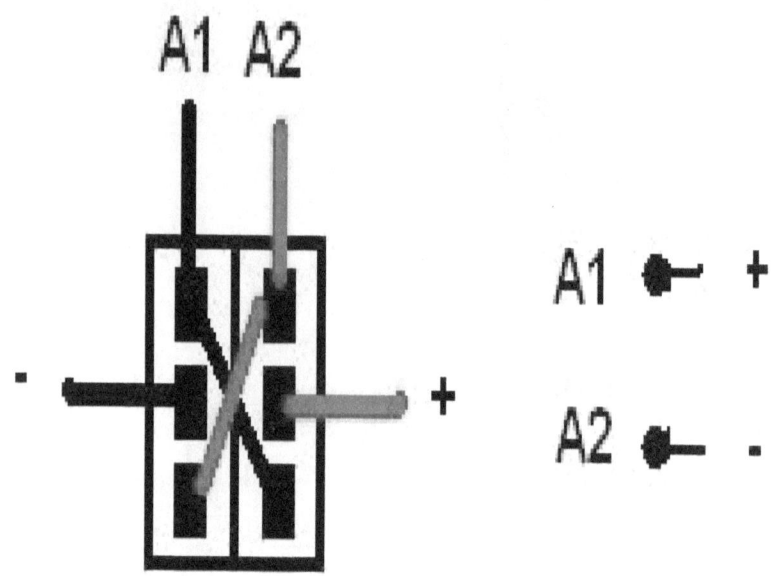

dc permanent magnet

center-off   on-off-on

to reverse interchange A1 and A2

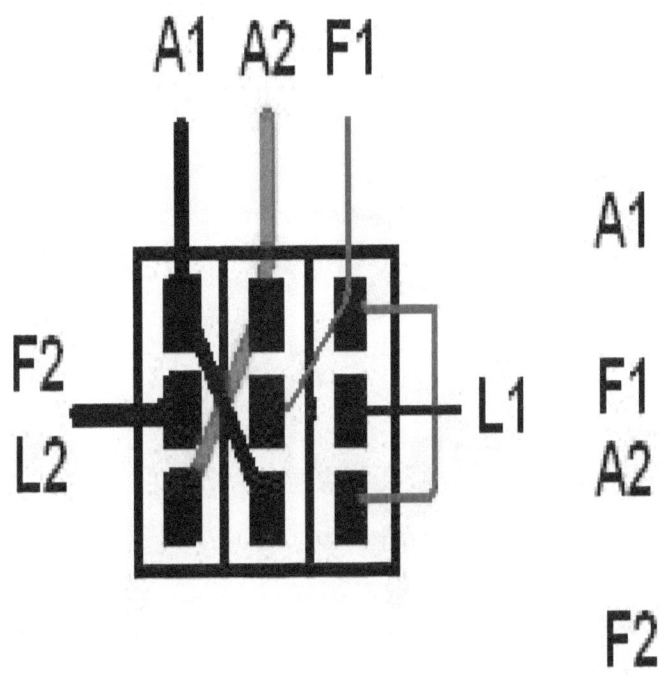

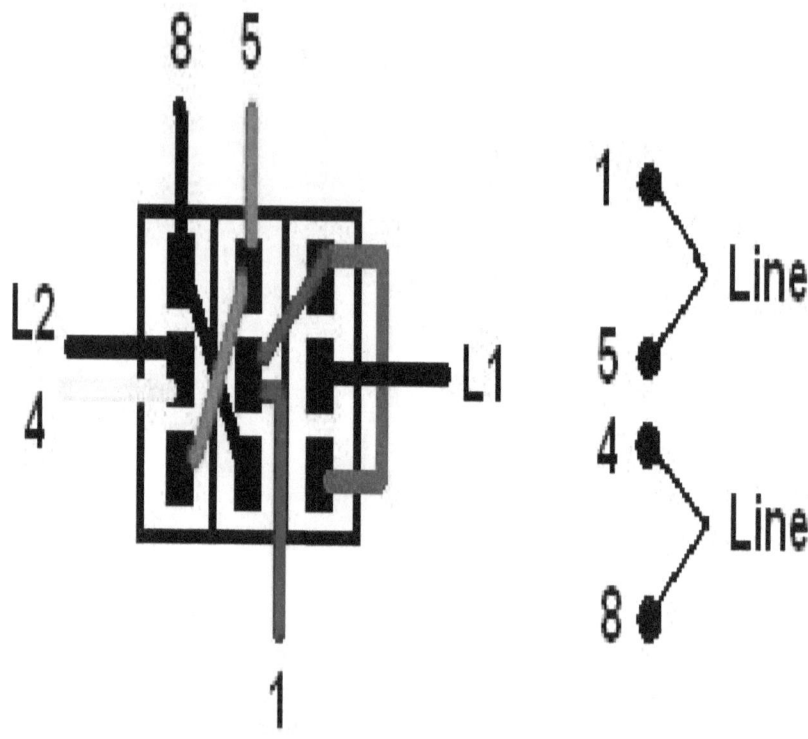

single phase   single voltage

center-off   on-off-on

to reverse interchange 5 and 8

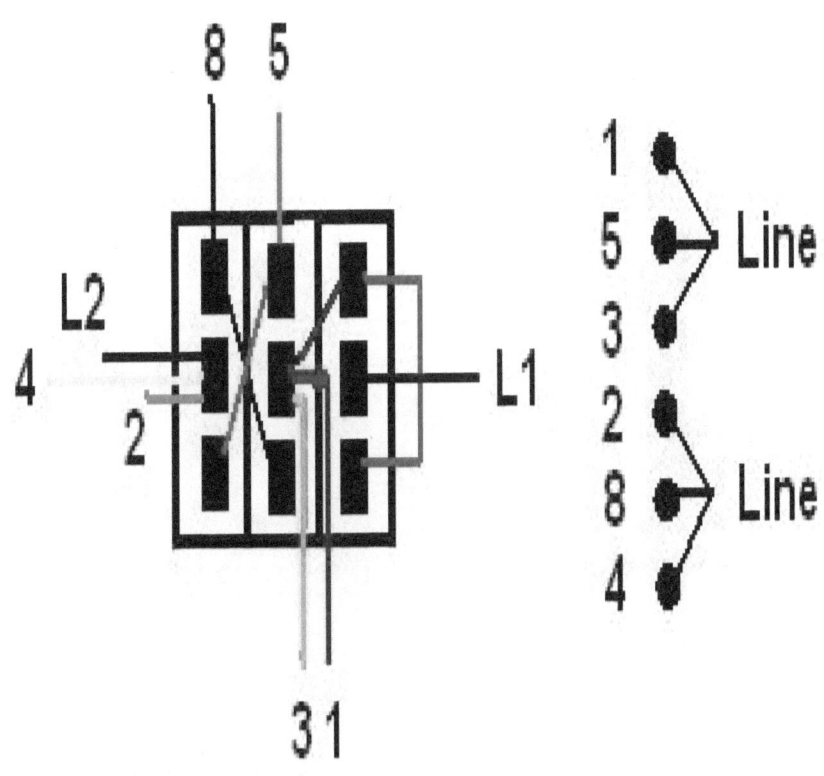

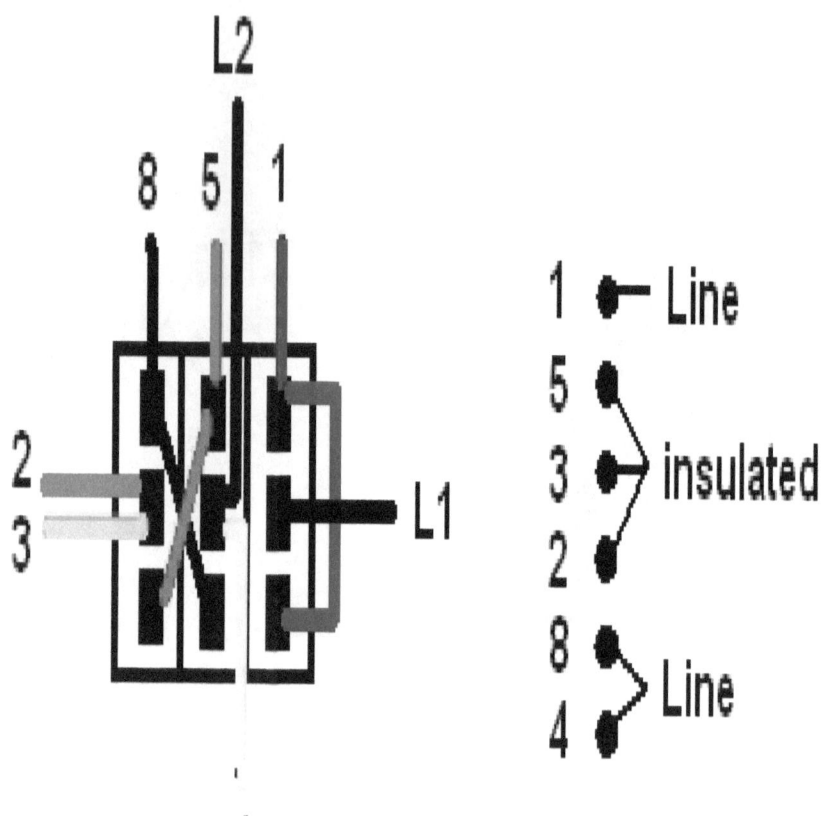

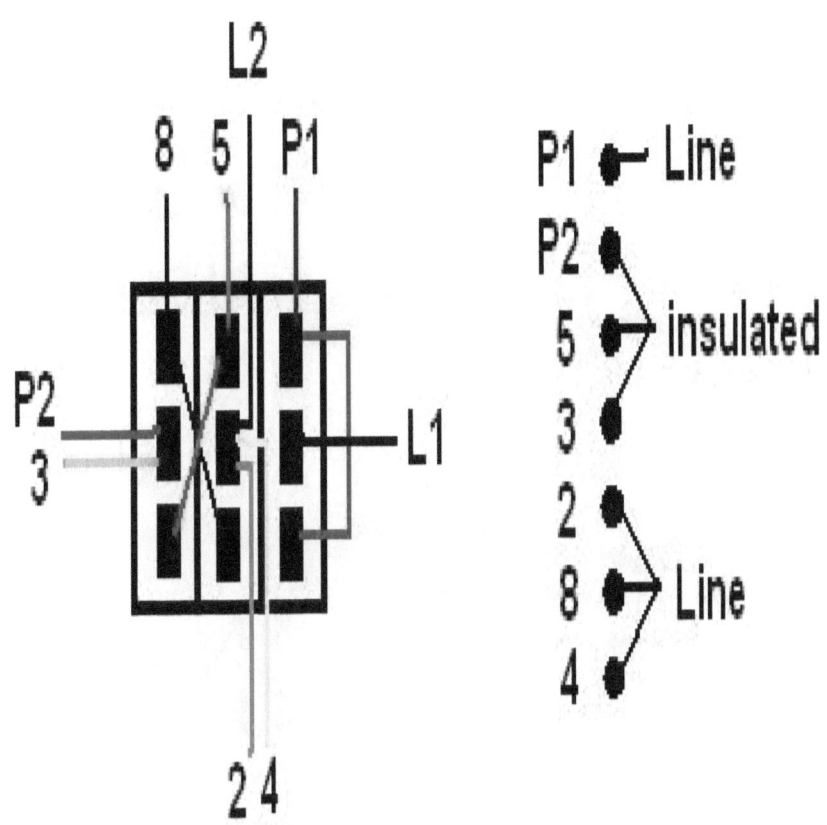

single phase low voltage thermal protection
center-off on-off-on
to reverse interchange 5 and 8

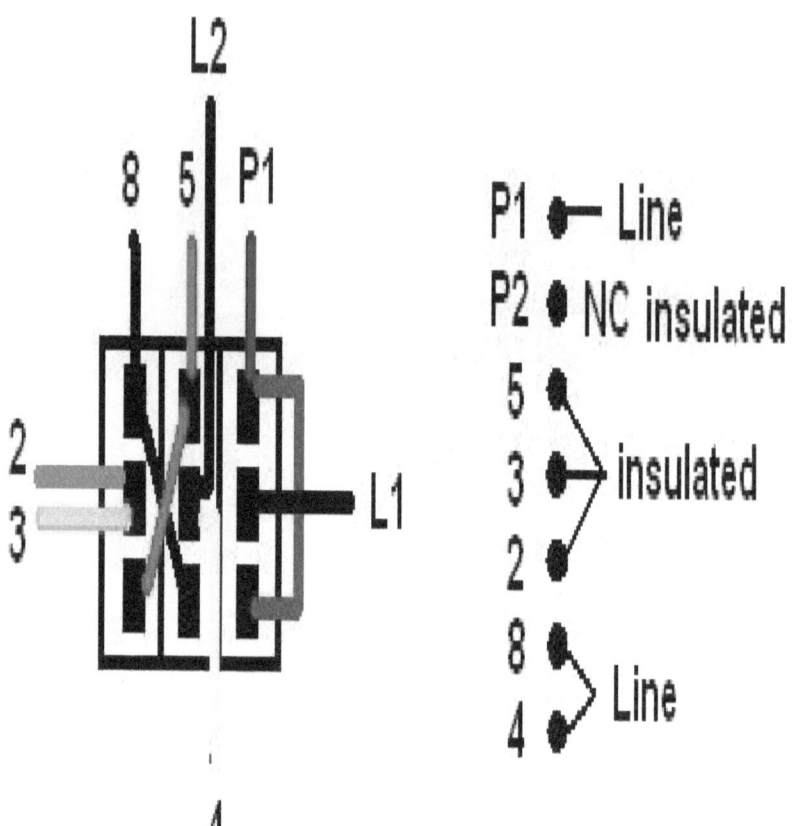

single phase   hi voltage thermal protection
center-off   on-off-on
to reverse interchange 5 and 8

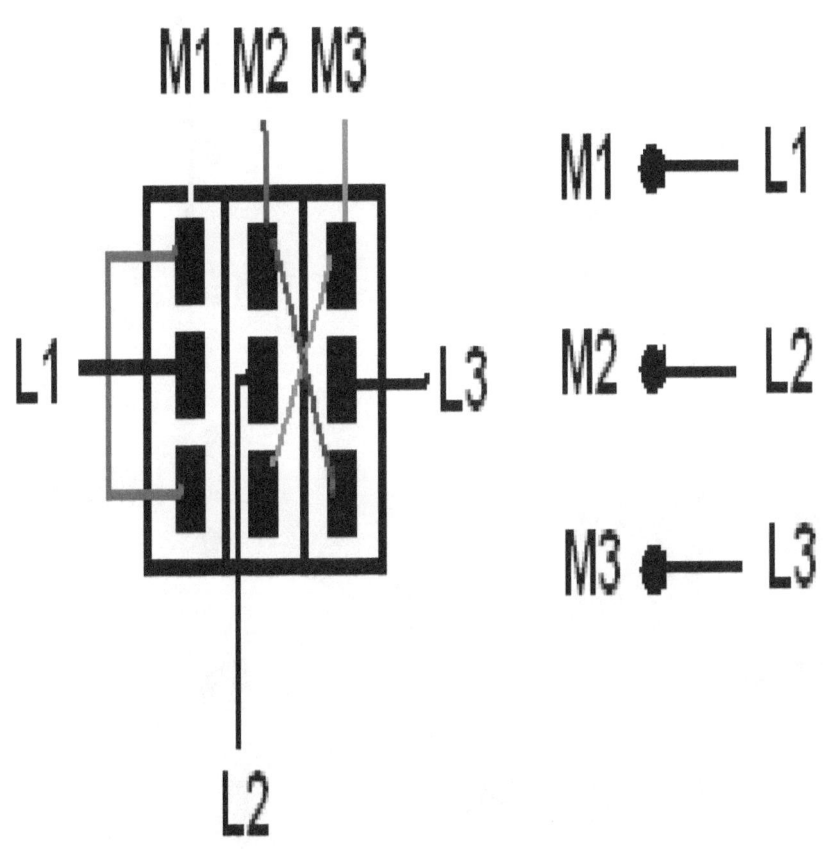

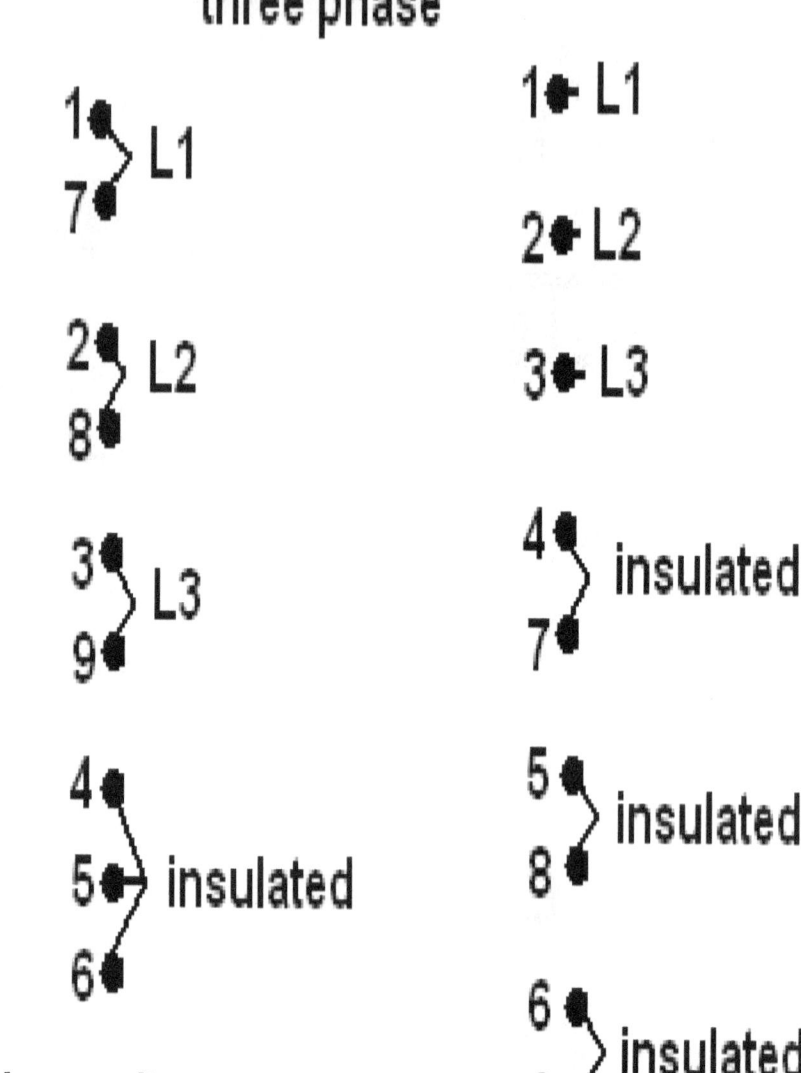

Drum switches are large heavy-duty manual switches. These types of switches are usually a 2-throw with a C/O (center-off) configuration. Drum switches usually control full current to a motor or device used for reversing applications. There are 2 basic types, the drum (slide contact) type or the contact (direct contact) type.

Common Drum switch types are:

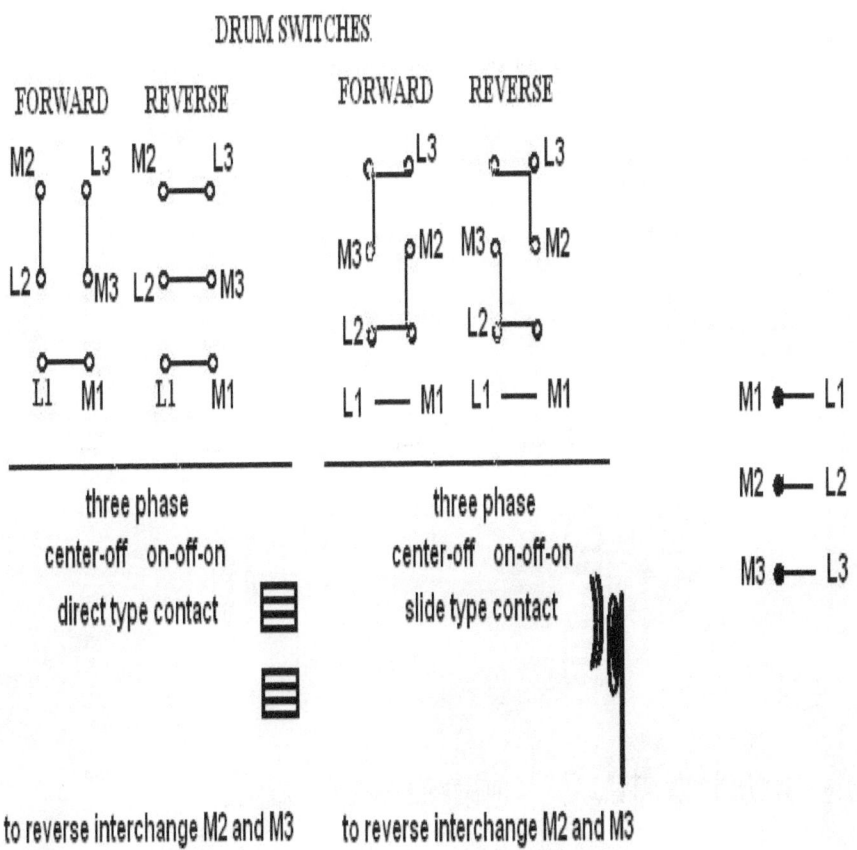

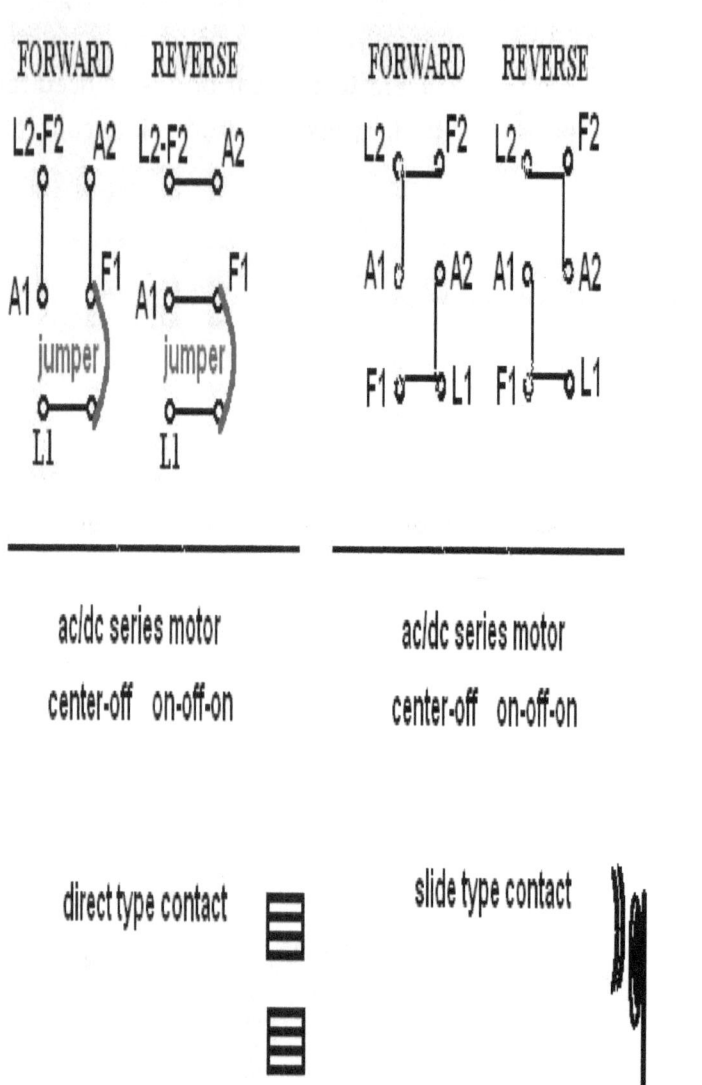

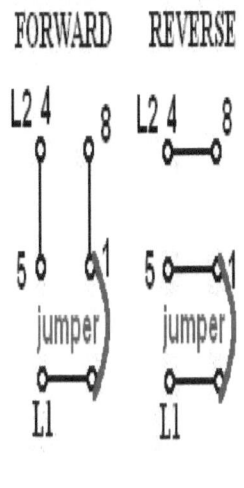
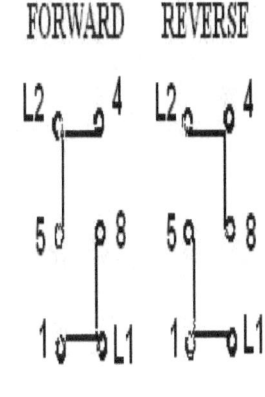

single phase   single voltage

center-off   on-off-on

single phase   single voltage

center-off   on-off-on

1 • \
5 •   Line

4 • \
8 •   Line

direct type contact

slide type contact

to reverse interchange 5 and 8   to reverse interchange 5 and 8

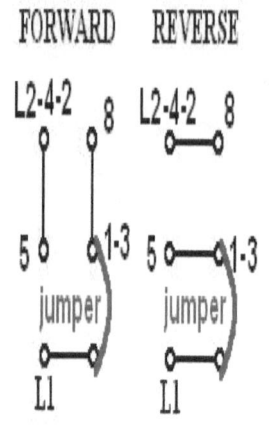

single phase   low voltage

center-off   on-off-on

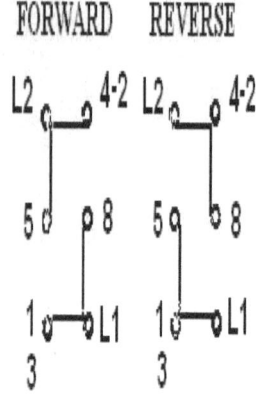

single phase   low voltage

center-off   on-off-on

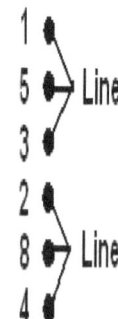

direct type contact

to reverse interchange 5 and 8

slide type contact

to reverse interchange 5 and 8

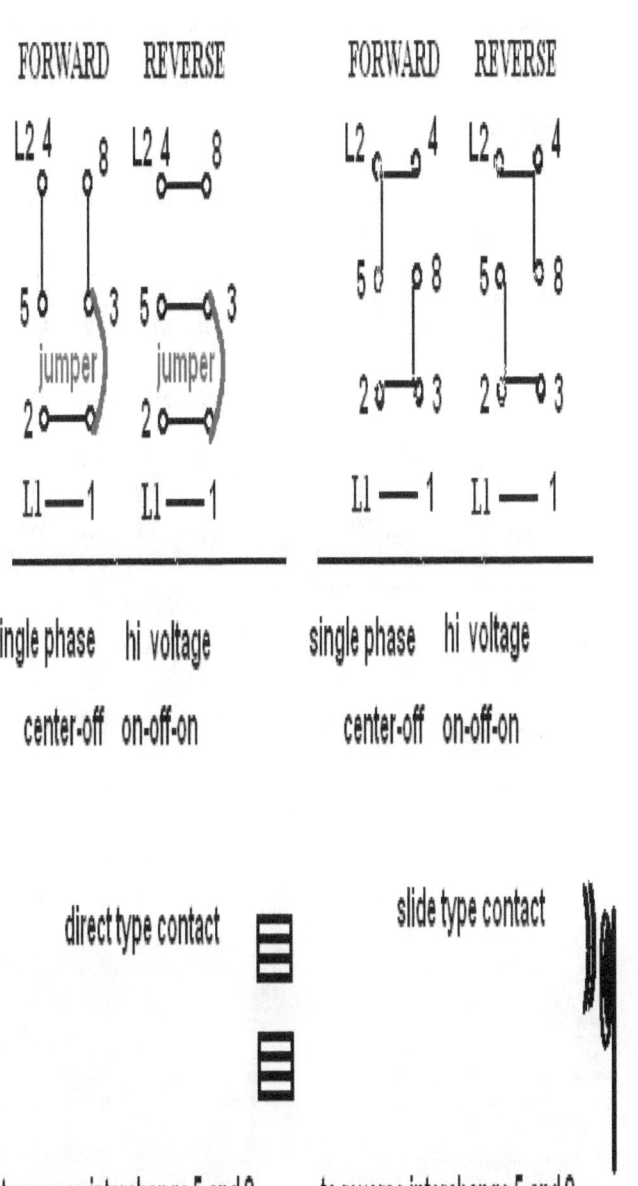

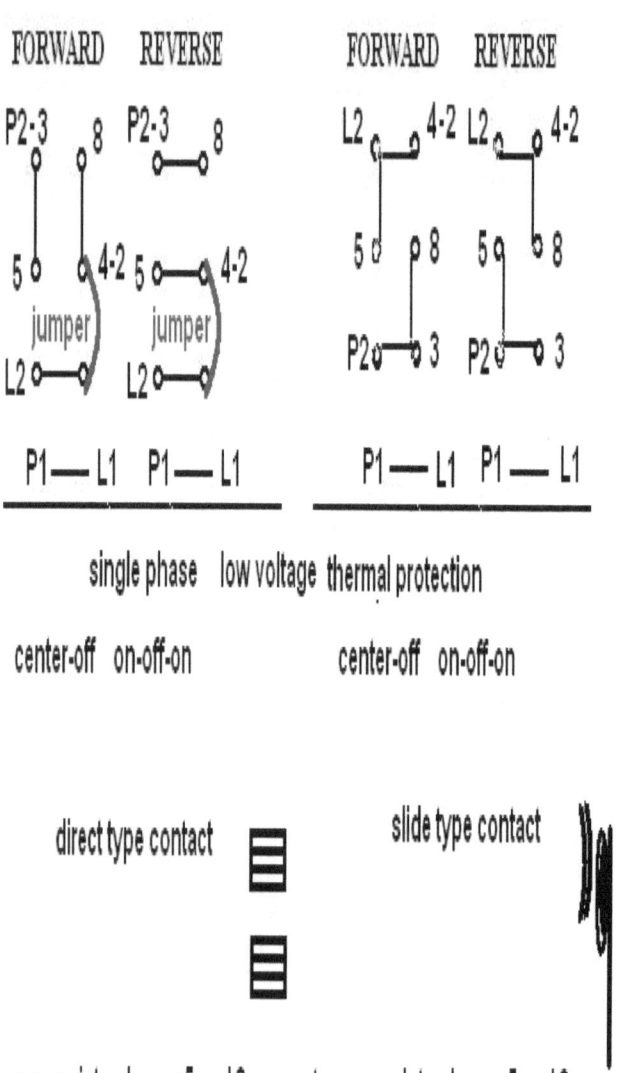

Rotary switches are used to distribute current to different contacts based on position and can have many active positions. These are not simple switches and come in a wide variety of types based on the Original Equipment Manufacturer (OEM). Some rotary switches may be as simple as a Single Pole Single Throw (position) on-off switch to multiple poles with greater than 20-Throws (positions). You should always refer to the manufactures connection with these switches.

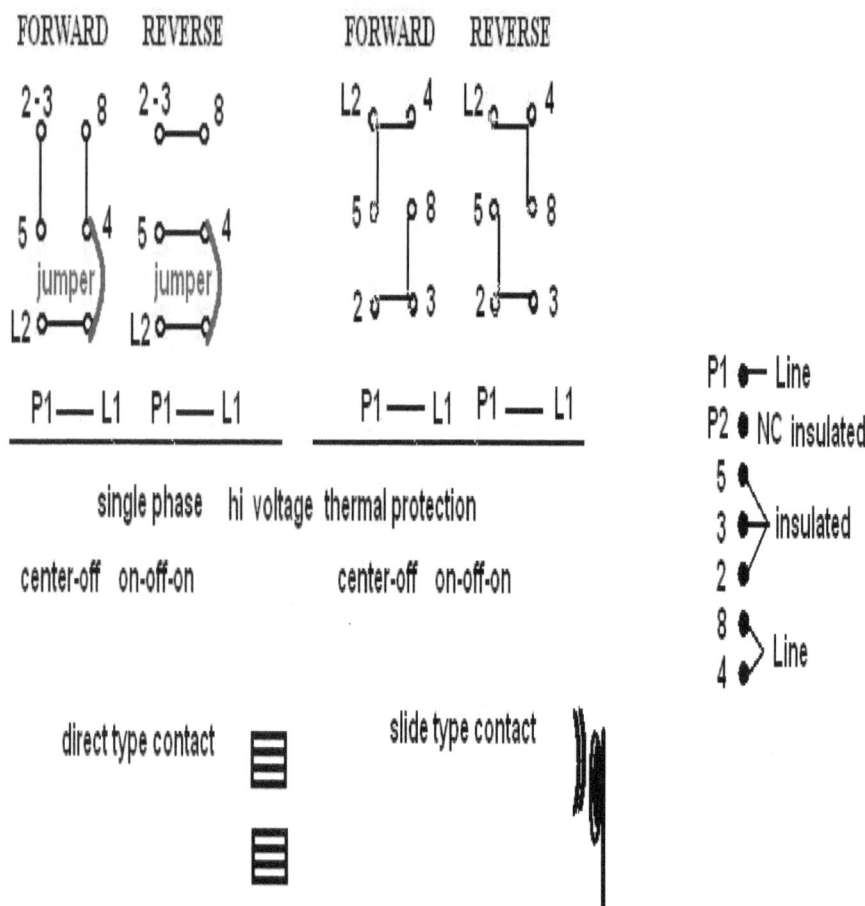

Slide switches also come in a variety of types from Single Pole Single Throw to multiple poles and multiple throws (positions). You should always refer to the manufactures connection with these switches.

Electrical mechanical switches sometimes called contactors, starters, solenoids, or relays are used for remote control of equipment. Contactors, starters, and solenoids are usually for high current applications while relays (control relays – CR) are used for low current applications. These switches come in a broad variety of poles with normally open, normally closed, or both contacts. They are usually Multi-Pole and Single or Double Throw types using Normally Closed (N.C.) or Normally Open (N.O.) contacts. Double throw types use two coils which control the on–off state and sometimes have mechanical inner-locks (safe operation) to allow only one throw or position during operation. Coils may be either laminated steel or solenoid types and come in a variety of voltages. The main part of this switch controls the full load amperage (FLA) to the device. These switches can also have multiple control contacts called auxiliary contacts or inner-lock (safe operation) contacts used for low current control. Contactors, starters, and solenoids may also have thermal protection sensors to disconnect the power supply in over-loaded conditions. The current rating is selected based on the load.

Common electrical mechanical switch types are:

single pole	double pole	three pole	three pole
=	= =	= = =	= = Z
N.O.	N.O. N.O.	N.O. N.O. N.O.	N.O. N.O. N.C.
⌇ coil	⌇ coil	⌇ coil	⌇ coil

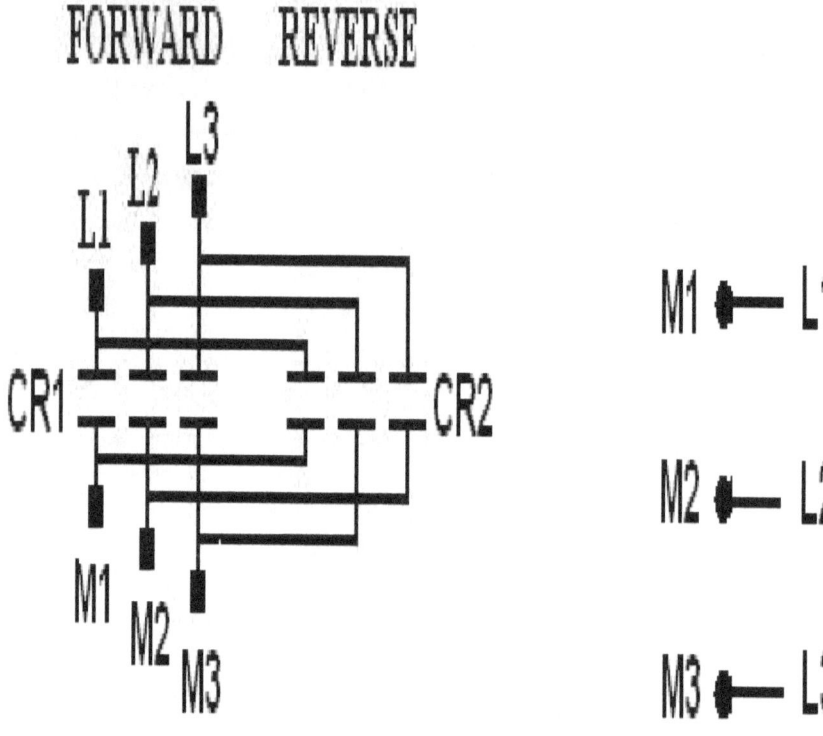

REVERSING CONTACTOR

FORWARD **REVERSE**

jumper L2 - F2

L1

CR1 CR2

F1 A1 A2

ac/dc series motor

3 - pole N.O. contacts

to reverse interchange A1 and A2

REVERSING CONTACTOR

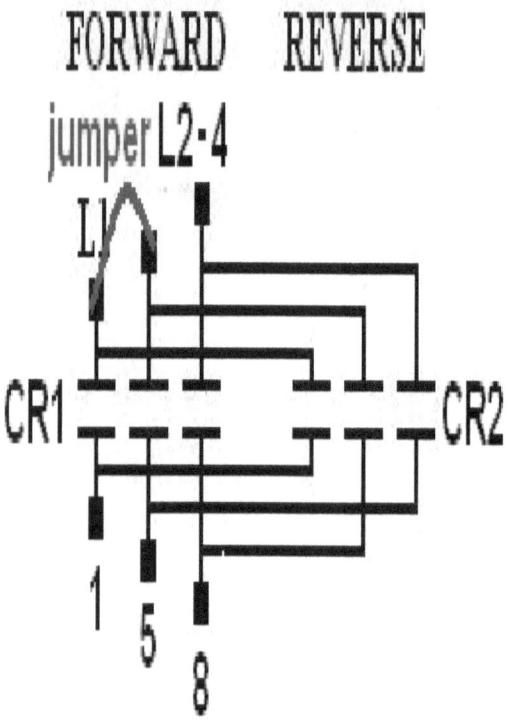
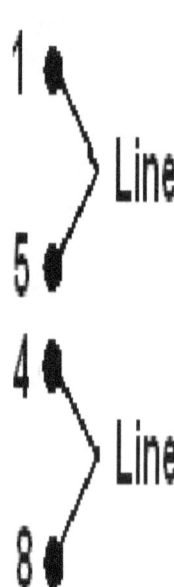

single phase   single voltage

3 - pole N.O. contacts

to reverse interchange 5 and 8

REVERSING CONTACTOR

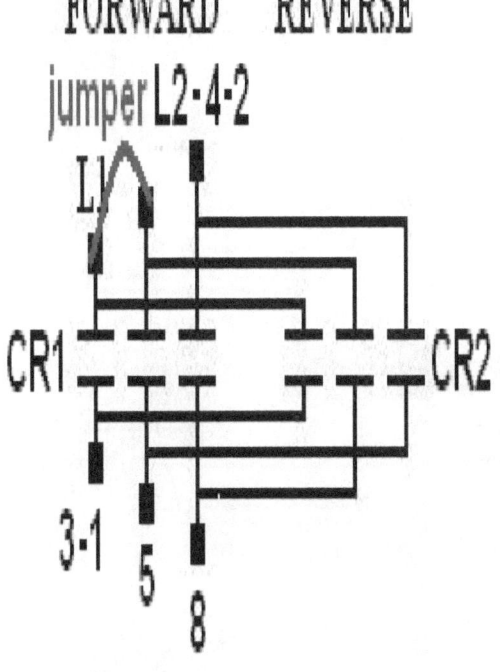

single phase   low voltage

3 - pole N.O. contacts

to reverse interchange 5 and 8

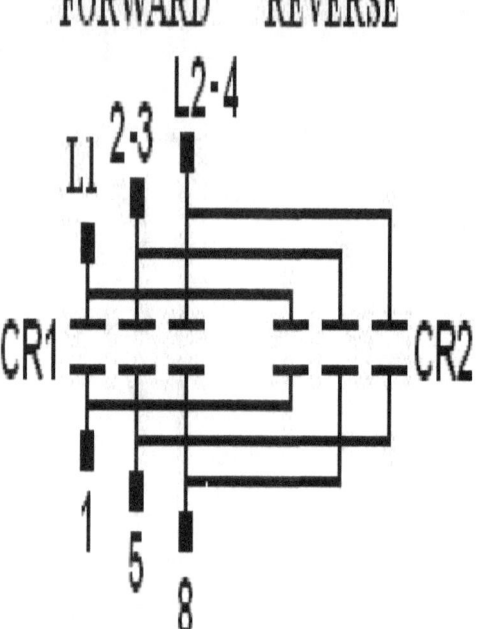

REVERSING CONTACTOR

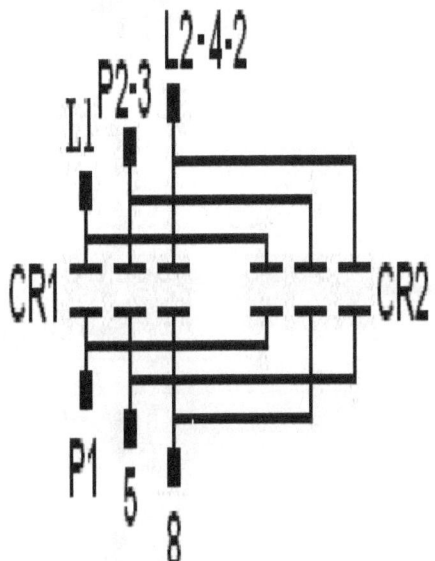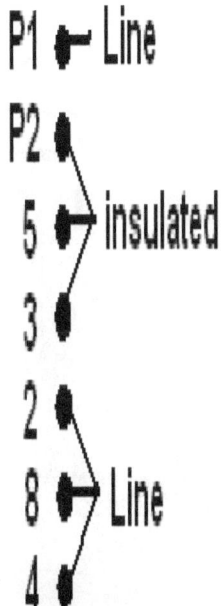

single phase   low voltage   thermal protection

3 - pole N.O. contacts

to reverse interchange 5 and 8

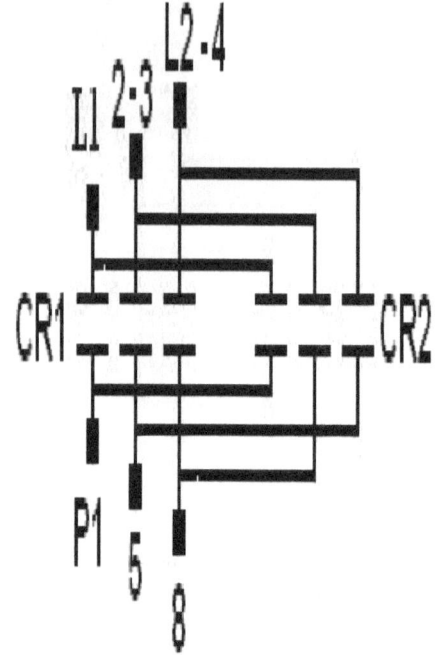 

single phase   hi voltage   thermal protection

3 - pole N.O. contacts

to reverse interchange 5 and 8

Other types of electric switches are used for speed control or light dimming. These two types are similar but not the same in operation.

The speed control switch uses an electronic alternating current (AC) device and usually controls 30% speed to Maximum speed adjustments. This switch can be used also in lighting applications. You should always refer to the manufactures connection with these switches.

The light dimmer switch uses an electronic direct current (DC) device that controls current from 0% to Maximum. This device cannot be used as speed control and will damage equipment if attempted. You should always refer to the manufactures connection with these switches.

Ladder Logic

Ladder Logic is used to map the way the control is achieved in electrical system operations. These are the steps or diagrams of operation used to control the operation of contactors, starters, solenoids, relays, and other sensing devices.

Common ladder logic diagrams and symbols are:

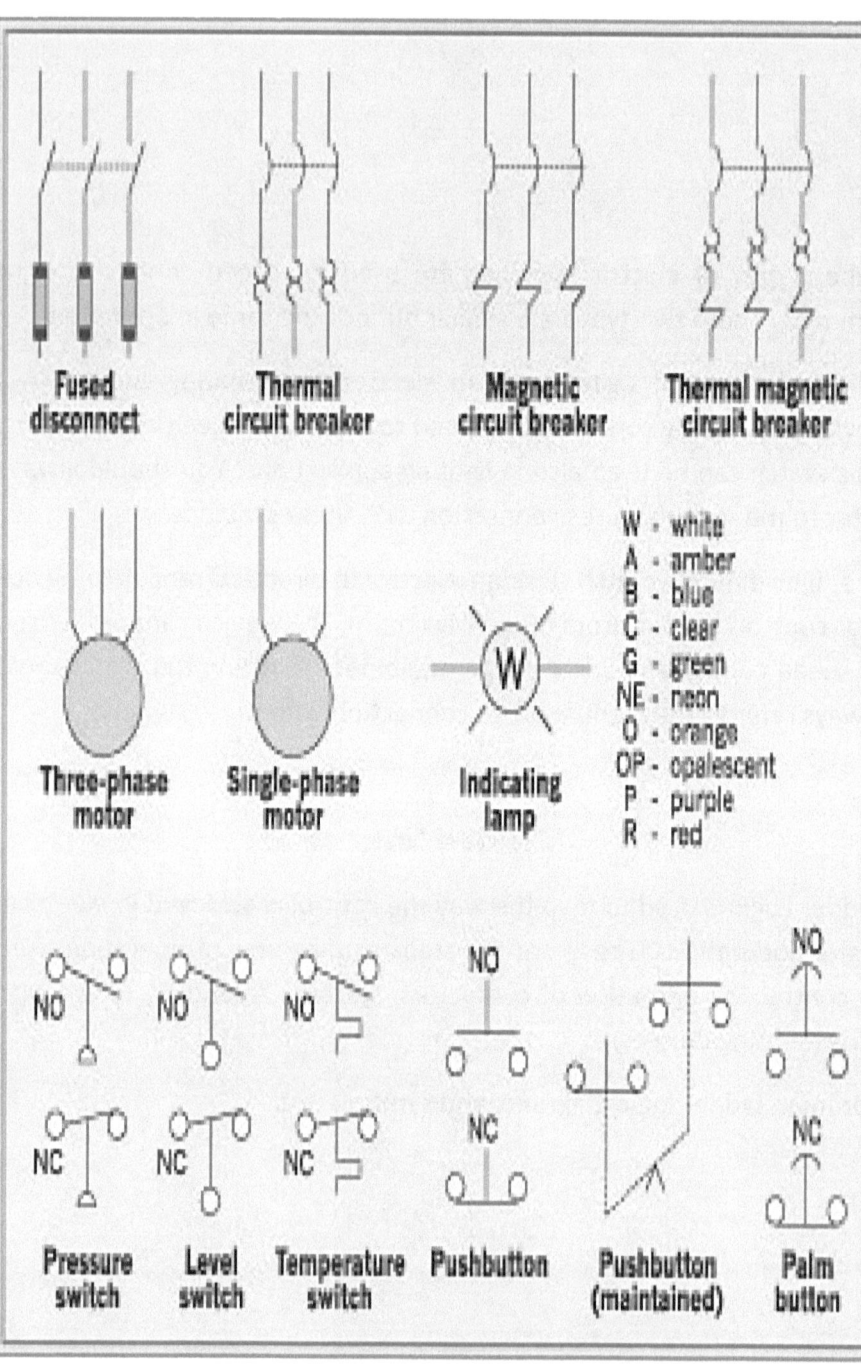

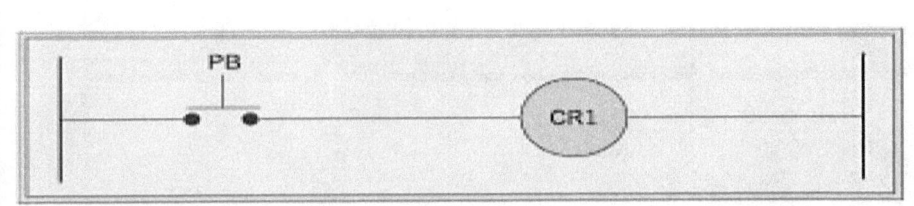

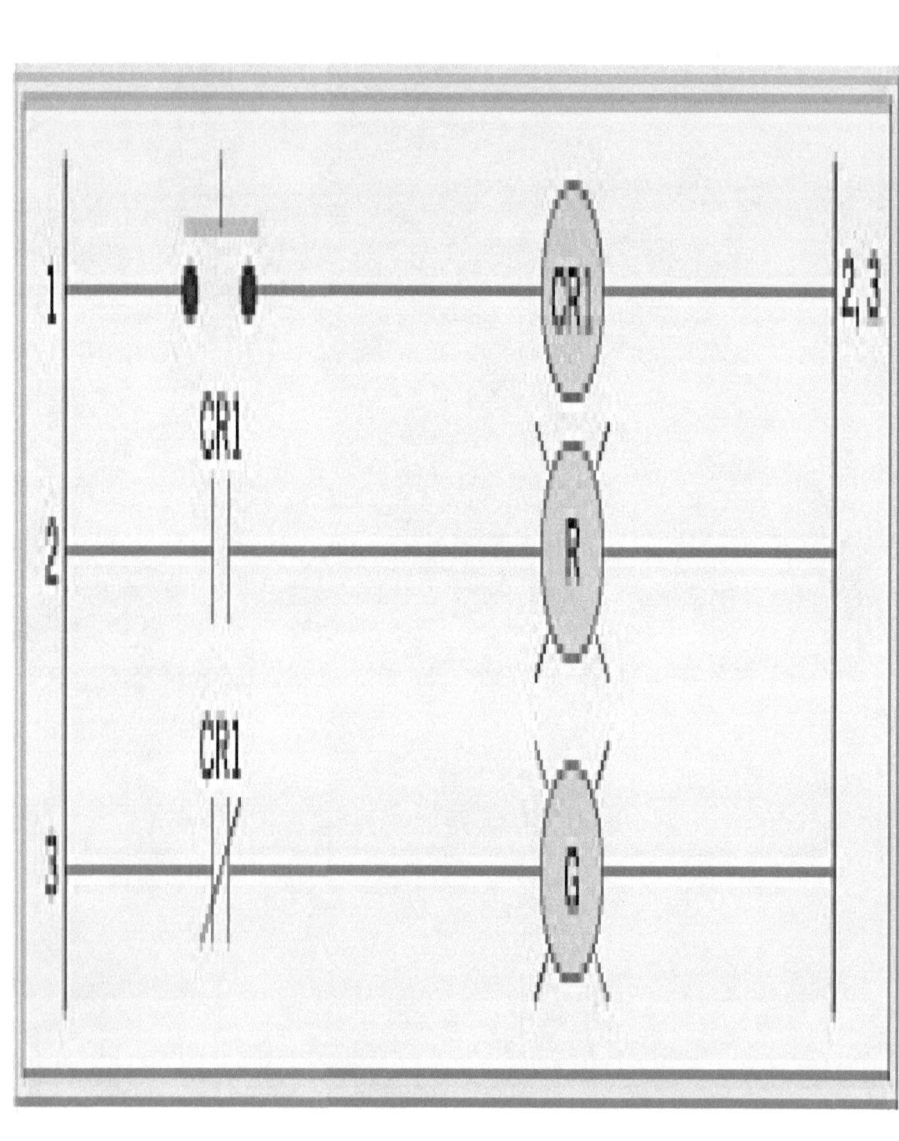

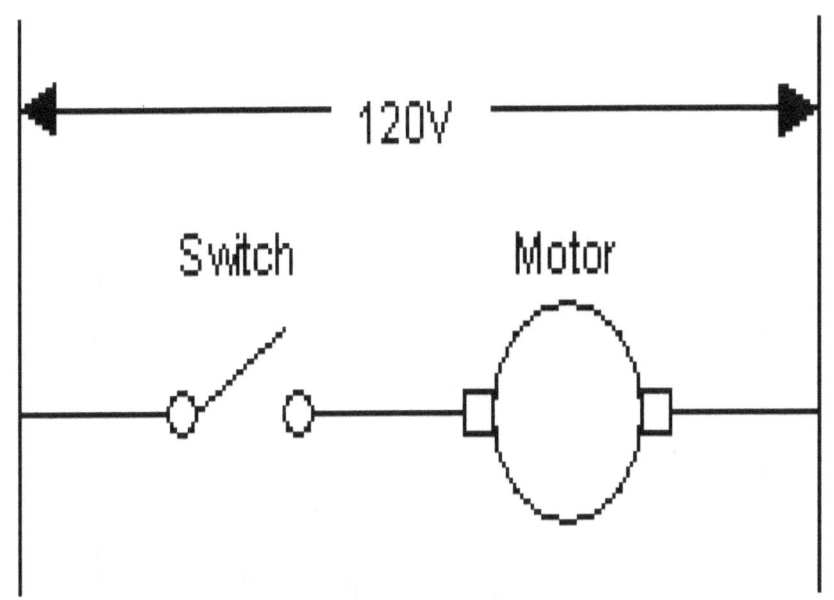

Electrical Diagram of a Motor and Switch - 1a

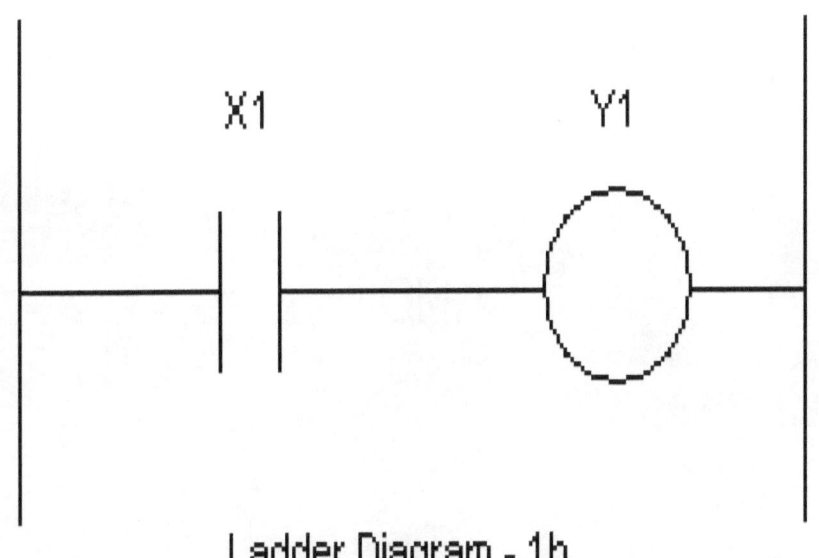

Ladder Diagram - 1b

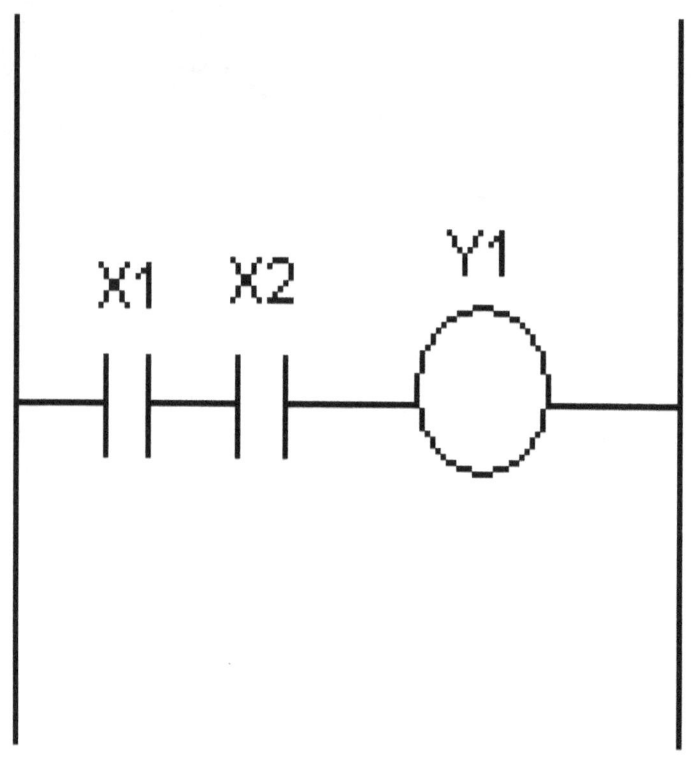

AND

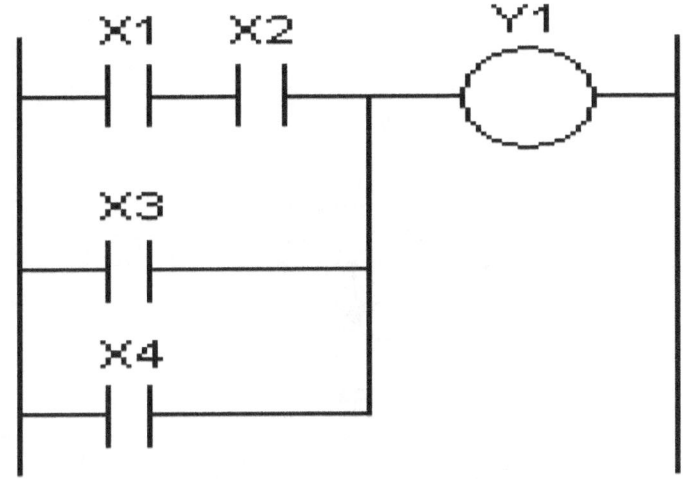

Combination

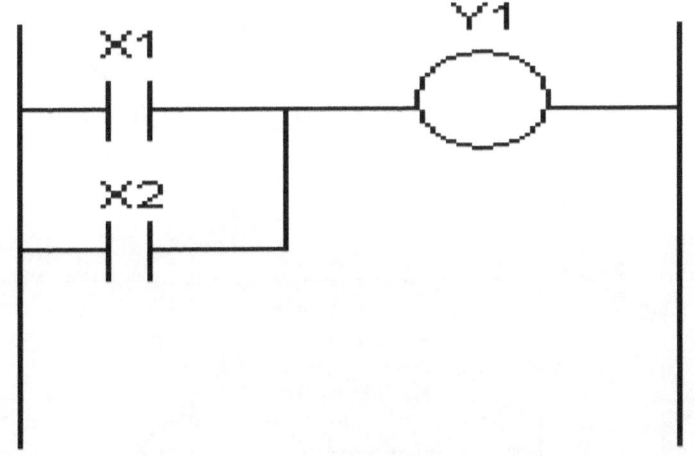

OR

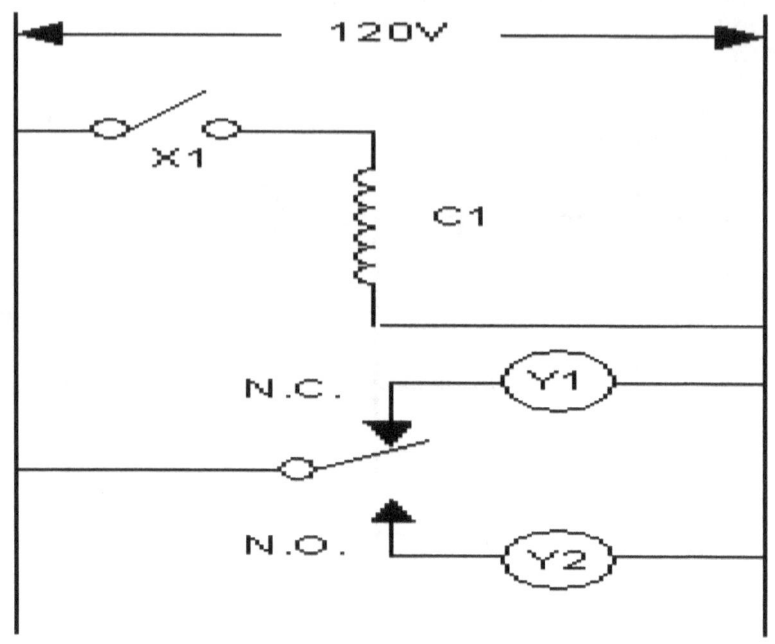

Electrical Diagram - 3a

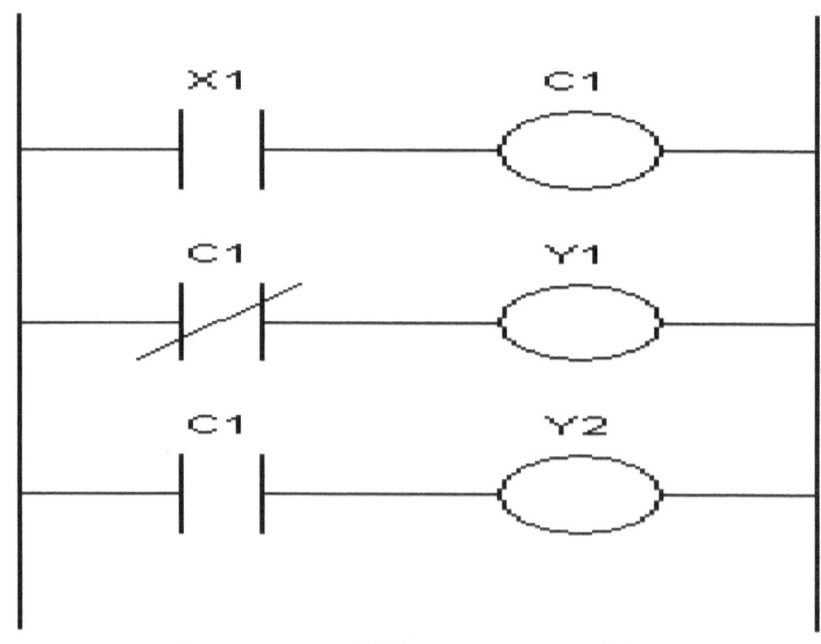

Ladder Diagram - 3b

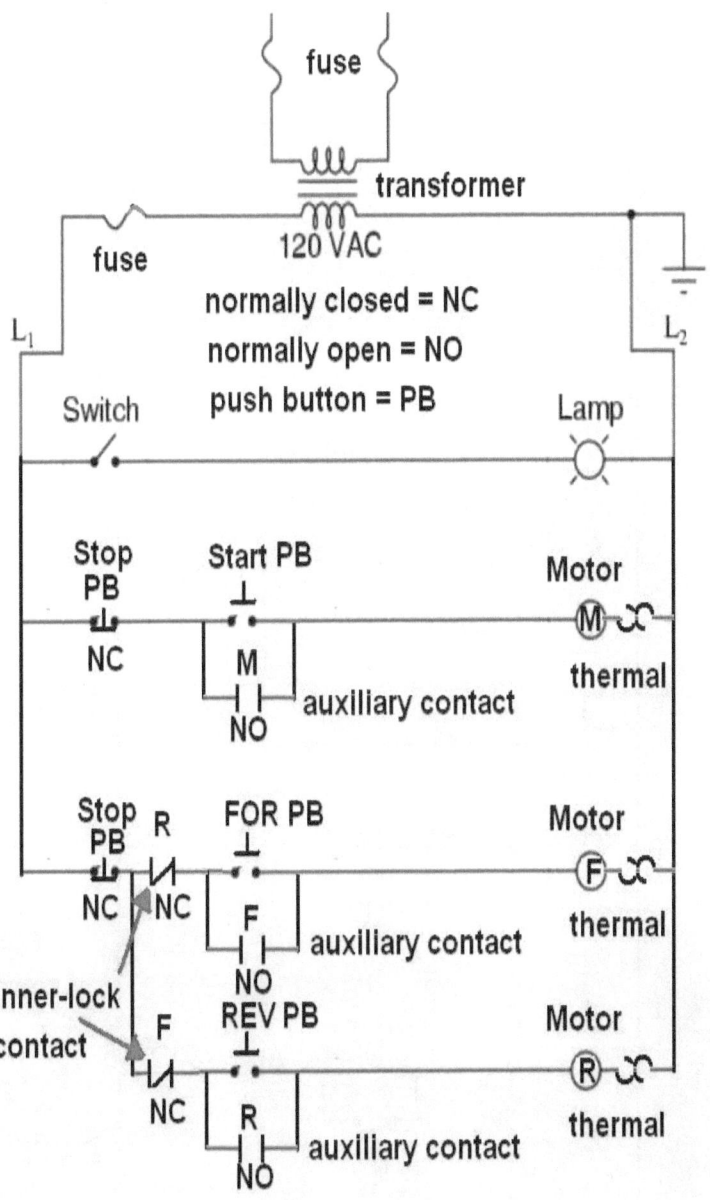

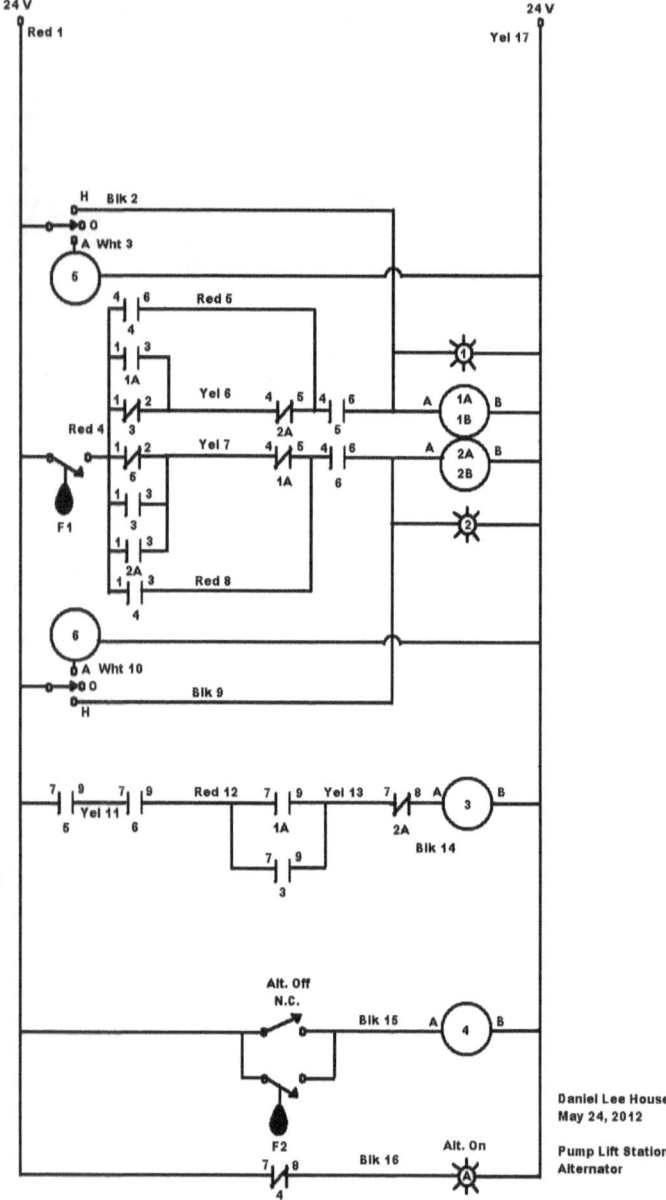

Each diagram displays a step-by-step operation or logic. Breaking a control circuit down into individual sequenced steps is needed to, not only, design a control circuit and build it, but to also troubleshoot it in the future.

To build a control circuit, start the diagram at the top, with the power supply, then the first operation, followed by the next operation, and so on. Keep the power supply leads to each side of the diagram respectively. With a little practice, this is easy. You should be able to read each process. The symbols you choose can be NEMA standard or your own idea of a symbol. Keep the diagram organized and neat for easy reading. You may also start at the top of a logic diagram and number each branch circuit sequentially. The lowest number (1) starts at the top power supply on the left, while the highest number (17) represents the other side of the power supply on the right, and also the highest number of circuits in the diagram (see diagram above – Pump Lift Station Alternator). This makes reading easier in complex control diagrams.

Phase Converters

About Rotary Phase Converters

Rotary Phase Converters are designed to generate three phase power from a standard single phase 230 volt line. This type of Power Converter is used to maintain three phase power to an entire system of three phase motors while eliminating the need of installing Static Phase Converters on each piece of equipment. The Rotary Converter consists of a three phase motor used as a generator and motor running capacitors (oilytic type) which are used to balance the output phases of the generator motor. Installation requires that the Rotary Converter be connected to a main disconnect panel rated for the total output current of the unit. Branch circuits are then taken off of the main disconnect and wired to the individual three phase disconnects to provide sufficient overload protection for each branch circuit. This unit is capable of starting motors as large as the generator motor and also supplying as much as twice the generator motor output power. An example would be that of a 15 H.P. Rotary Converter being installed, the maximum starting power of this converter would be 15 H.P. while the maximum total combined output power would be 30 H.P. This means that the largest motor that is required on the system determines the smallest practical generator motor used. The rest of the total combined horsepower load should not exceed twice the generator motor size, or you should use a larger generator motor. These units are sized by output power requirements and at no time should be used to start motors that exceed the input generator motor size. Industrial three phase wiring standards should be maintained to provide safe distribution of power between loads. There may be an initial expense in wiring with this method but the overall output three phase power provided will more than pay for itself. When properly sized and installed the Rotary Converter will last for many years of satisfactory service. To adjust the produced phase voltage down, remove just enough capacitance (1 capacitor at a time) while fully loaded, this capacitance may be required again in the future as load demand changes.

About Static Phase Converters

Static Phase Converters are designed to operate three phase motors on standard single phase 230 volt power. This type of Power Converter uses motor starting capacitors (electrolytic), motor running capacitors (oilytic), and a voltage sensitive relay in conjunction with the three phase motor winding to generate usable three phase power. On larger units a contactor is used with the voltage sensitive relay due to limited current handling capabilities of the voltage sensitive relay. Both smaller and larger units are connected to the motor leads with three output leads marked C1, C2, and C3 respectively. This makes connecting the converter to a three phase motor a relatively easy task. All that is needed is to connect each of the converter leads across the motor leads, T1,T2, and T3, and supply 230 volt single phase power to C1-T1 and C2-T2 with some type of disconnecting means. C3-T3 connection is hard wired and insulated. A common problem is when the converter is not connected properly to the motor in use. When this happens the motor will not operate correctly and may result in damage to the motor, converter, or both. Remember that the converter uses the three phase motor to aide in the developing of three phase power. To reverse rotation on the motor, interchange any two motor leads as you do a standard three phase motor. Some units are equipped with an external indicator lamp which is wired internally to the starting capacitors to indicate that the motor is operating correctly. This lamp should turn on upon starting of the motor and should turn back to off after about 5 to 10 seconds, depending on how heavily the motor is loaded. If the lamp does not turn off it is an indication of improper wiring or an overloaded motor. When wired correctly to the proper size motor, this Static Phase Converter will provide many years of satisfactory service.

Output Power Applications

On some types of equipment such as lathes and drill presses, when a motor runs idle until a load is applied or is lightly loaded, it is possible to obtain usable three phase power from the converter driven motor. Power is taken off by connecting tap leads to the converter leads C1, C2, and C3. The converter driven motor must be started first in order to utilize this option. While the converter driven motor is running idle, the three phase power generated is capable of starting other three phase motors not exceeding one half of the horsepower rating of the converter driven motor. Also the power generated by this means is capable of running a combination of motors not exceeding the converter motors nameplate horsepower rating. An example would be that the converter driven motor is 2 H.P., while the 2 H.P. converter motor is running idle, it is possible to start two 1 H.P. motors or a total of four 1/2 H.P. motors off of the 2 H.P. converter driven motor. Care should be taken to never exceed the maximum starting horsepower or the maximum combined horsepower ratings and to not load the converter driven motor when used in this manner. Never attempt to use this method on any converter driven motor that has a heavy load applied or damage may result to the loaded motor, converter, and /or the added motors.

Phase Converters Are Here To Stay

Many years ago it came to my attention that you could operate three phase motors on a single phase power supply. For you readers not so familiar with the technology of phases, this simply means that a motor designed to be operated for industrial uses, can now be operated in your garage workshop or basement using household power supplies and provide the same economical power savings without expensive power service from the utility company. The way this is done is by installing a phase converter on the equipment to be operated. The phase converter converts standard single phase power into industrial three phase power thus allowing you to operate an industry classed machine from a simple power source. The art of building phase converters is nothing new, however having experience with types available on the market today, and the flaws in these converter designs, I would like to simplify the design process so that the common hobby achiever can understand how to build the best converter that money can buy. If you are interested in installing any type of industrial equipment in your workshop or business and you have access to a Refrigeration and Electrical Wholesaler, you have the means to acquire the parts needed to build a phase converter. I learned in college to K.I.S.S. (keep it simple stupid), so that you can achieve success by following these simple instructions. I will explain the basics of three phase motor operation and guide you through the designing and building of your own phase converter so that you can take advantage of the savings, both in the manufacture of your own phase converter and the purchase of industrial equipment. These phase converters are useful on any motor driven equipment such as air compressors, air conditioning compressors, lathes, grinders, mills, drill presses, saws, conveyors, pumps, and many other applications where industrial power is used.

About Three Phase Motors

Three phase motors are designed to give greater efficiency for the manufacturing industry by providing low cost of operation through less current draw than typical single phase motors and less maintenance through fewer moving parts. Single phase motors, in other words, provide no efficiency and consume more current, sometimes as much as twice the current of three phase motors, and are more likely to fail due to more moving parts. Without getting into complex analysis of each type of motor you can already see why three phase motors are the working medium for industry. The only problem is that three phase motors require three power leads and your standard power around your home and shop is only two power leads. If you try to operate a three phase motor on single phase power without some means to provide power for the third lead the motor will overheat under load and short circuit, thus damaging the motor. A phase converter supplies the third lead needed for the motor to operate correctly.

Purchasing Your Equipment

Due to continued changes in industry today, many machines are becoming obsolete and scrapped while being replaced with newer and faster machines. In most cases, these older machines are still in operating condition but just can't keep up with the demand of today's production requirements. In many instances manufacturers are selling these good machines at sales and auctions while closing current production lines in order to manufacture parts in other areas of the country. You can take advantage of these situations by acquiring these machines and operating them with a phase converter.

Background on Phase Converters

As a computer science and electrical engineer and designer, I have created a simple method of constructing a phase converter from locally available components. This design of converter has built in to it all the good things that other converters on the market don't have and thus the converter sold on the market many times does not last and fails. There are two basic types of phase converters; the first type being called a static type converter. The static converter is normally used to start one piece of equipment so that each machine would require a converter. If you have many machines, this could run up into large costs just for the converters. When many machines are used, the second type converter is best used being called a rotary generator converter. This type of converter is actually a three phase motor used to generate three phase power from the incoming single phase power. The terms you should understand are Horsepower - the rated output working power of a motor, RPM - the rated revolutions per minute of a motor, Voltage - the rated operating voltage of a motor, Amp - the rated full load operating current of a motor, Potential relay - an electrical mechanical device that opens a set of contact points at a sensed voltage level, Contactor - an electrical mechanical device that closes a set of contact points to supply power to a load, Capacitor - a component device used to cause a phase shift in starting and maintaining power balances, Discharge resistor - a component device used to bleed off power from a charged capacitor. Be sure to follow the requirements for these devices and components when building a phase converter and the converter will provide you with years of service.

Selecting Designs and Enclosures

Static converters are designed two different ways depending on the size motor that you intend to use. Small motors up to 5 horsepower use wiring diagram for method 1. Static converters from 7.5 horsepower to 15 horsepower use wiring diagram method 2 design (* use this formula to select the contactor for method 2 designs, (3 X (FLA)/contactor poles = Amp class of contactor). Rotary converters are also designed in two different ways. Most rotary three phase motors will start using wiring diagram method 3 but due to differences in spans and rotor bar makeup of the motor design itself some rotary motor designs need to be wired according to method 1 adding * 45 micro farads of phase shift (electrolytic capacitance) per horsepower of rotary generator design. Whether you are building a static or rotary design, you should always allow enough room in the enclosure to mount all of the components without cramming space. As a rule of thumb allow at least 2 inches from any live power source and a frame and always use a ground wire within the enclosure. Refer to the starter, wire size charts, and protective device diagrams for help in selecting controls for your phase converter. Always follow Starter Protective Device Charts or Formulas for correct wire sizes for line lead sizes.

* Example for selecting contactor for method 2

3 (20 fla) = 20 amp contactor @ 3 poles

3 poles

* Example for selecting starting capacitance for rotary unit

5 horsepower motor times 45 microfarads = 225 microfarads

Designing a Static Converter

For a static phase converter design type, be sure that the motor you intend to use is 230 volt connectable. You will need to obtain an enclosure large enough for the following parts:

1- Potential starting relay rated at 2 - 5 horsepower at 230 volts with a coil resistance of approximately 10 thousand ohms, commonly used for air conditioning compressors and submersible water pumps. This part is available from a refrigeration dealer, a pump dealer, or an electric motor shop. The two most popular brands are RBM/ESSEX or GE/MARS. This is a three terminal device labeled 1, 2, and 5 respectively (see Potential Relay Parameter sheet).

5 feet of insulated motor lead wire - 14 gauge AWG for up to 5 horsepower method 1 and 12 gauge AWG for 7.5 to 15 horsepower method 2 designs.

1 - General Purpose Contactor, 2 - pole, 230 volt coil, rated at 40 amp for method 2 designs.

Electrolytic type capacitor(s) connected to obtain the calculated requirements rated at least 250 volts using the software computer calculator or the following calculation method:

Calculate - Multiply the full load amp rating of the motor being used times 33 to obtain the correct micro farad size for stating torque. Attach a 15 thousand ohm 3 watt discharge resistor across the terminals of each capacitor.

Oilytic type capacitor(s) connected to obtain the calculated requirements rated at least 240 volts using the software computer calculator or the following calculation method:

Calculate - Multiply the full load amp rating of the motor being used times 10.5 to obtain the correct micro farad size for running. Attach a 100 thousand ohm 1/2 watt discharge resistor across the terminals of each capacitor.

Rosin core solder or several 1/4 inch female spade connectors - for connecting capacitors, resistors, and relay.

Several 1/4 - 20 machine screws, nuts, washers, and plumbers strap - for securing capacitors in the enclosure.

Connect components and capacitors according to method 1 or method 2 diagrams.

Designing a Rotary Phase Converter

For a rotary phase converter type design you will need to obtain the following parts:

1 - Three phase, 230 volt, 1800 RPM, motor to act as a generator. This motor must be as large as the biggest motor you want to start on the three phase converter system. If you plan to start a 15

horsepower motor on your equipment, you must have at least a 15 horsepower rotary generator to build your system. The rotary generator motor will never have a mechanical load attached to the motor shaft and will serve as an electrical generator for the three phase electrical system. The combined output power of this type of converter is twice the rated horsepower of the generator motor used. In other words, a 15 horsepower generator motor will operate any combination of motors totaling 30 horsepower and none of the motors exceeding 15 horsepower.

Oilytic type capacitor(s) connected to obtain the calculated requirements rated at least 280 volts (**for 460 volt rotary converters, connect capacitors for at least 500 volt rating and use 460 volt current rating on the rotary motor to calculate capacitance using the software computer calculator. This capacitance will be ½ that of the capacitance for a 230 volt rotary motor of the same size.**) Using the software computer calculator or the following calculation method:

Calculate - Multiply the full load amp rating of the motor being used times 11.5 to obtain the correct micro farad size for running. Attach a 100 thousand ohm 1/2 watt discharge resistor across the terminals of each capacitor.

Several 1/4 - 20 machine screws, nuts, washers, and plumbers strap - for securing capacitors in the enclosure.

Connect components and capacitors according to method 3 diagram, note, **if generator converter motor does not start** with method 3, add 45 micro farads of electrolytic capacitance per horsepower of generator motor and connect according to method 1 diagram. The reason for this is dependent on the span of winding coils and the number of rotor bars used in the motor generator, it has nothing to do with manufacturer.

After you connect your load to the rotary converter, measure the voltage across the yellow or produced line leg of the converter. This voltage can be adjusted down by reducing capacitors gradually. This can only be adjusted after the load is applied; for example the more load on the, the converter, the more capacitors that are needed. The less load on the converter, the fewer amounts of capacitors. This voltage should not exceed 280 volts nor be less than 208 volts.

Formulas for Converters

$2 \times (3.1416) \times (60) \times (Capacitance) \times (230)$ = **Current for leads**

Or:

$86{,}707.957 \times (Capacitance)$ = Current for leads

Example:

10 MFD = 10 Micro Farads = .000010 Capacitance for calculations

.000010 = Capacitance

$86{,}707.957 \times (.000010)$ = .867 Amps

Capacitor lead sizes:

16 AWG = 10 Amps (smallest size for mechanical strength)

14 AWG = 15 Amps

12 AWG = 20 Amps

10 AWG = 30 Amps

8 AWG = 40 - 50 Amps

6 AWG = 60 - 70 Amps

4 AWG = 80 - 100 Amps

Adding Capacitors Series:

$$\frac{1}{\frac{1}{\text{Capacitance 1}} + \frac{1}{\text{Capacitance 2}} + \frac{1}{\text{Capacitance 3}} + ...}$$

Series Capacitance doubles the voltage capacity

Adding Capacitors Parallel:

$$\text{Capacitance 1} + \text{Capacitance 2} + \text{Capacitance 3} + ...$$

Parallel Capacitance adds the Capacitance with the voltage the same

Cord Size for long cords over 30 Feet:

$$\frac{30 \, (\text{Volts})}{5 \, (\text{Amps}) \, (\text{Feet})} = \text{Resistance in Ohm's per 1000 Feet}$$

This can be translated to a wire size using the properties chart for copper wire.

Potential Relay Parameters

GE / MARS

Ideal Size   6.8K Ohm Coil Resistance 336V Continuous coil voltage

MARS 65 work best rated at 336V

Essex / RBM			GE
Coil #	Continuous Volts	Ohms Resistance	Coil #
1	130V	750	7
2	170V	1.4K	2
3	256V	3.3K	5
4	336V	5.2K	3
5	395V	7.1K	10
6	420V	8.2K	6
7	495V	12K	4

Example of Essex / RBM Models

128 | 11 | 6 | -23 | 4 | 2 | C

the second number from the right side is the coil number

in this case it is # 4

4 coils work best rated at 336V

Starter, Wire, and Protective Device Size Charts

Follow the Starter Size and Wire Size Chart below to determine proper size for each Branch Circuit.

NEMA Size Starters	Maximum Horsepower for Three Phase Motors
00	1 1/2
0	3
1	7 1/2
2	15
3	30
4	50
5	100
6	200

Percent of Full Load Motor Current

Type of Motor	Non-time Delay Fuse	Time Delay Fuse	Instant Breaker	Timed Breaker
no code	300	175	700	250
code F-V	300	175	700	250
code B-E	250	175	700	200
code A	150	150	700	150

Recommended Size Copper Wire AWG per Distance from Main Disconnect

100 ft	150 ft	200 ft	300 ft	HP
12	12	12	12	1 1/2
12	12	12	10	2
12	10	10	8	3
10	8	8	6	5
8	6	6	4	7 1/2
6	4	4	4	10
4	4	4	2	15
4	2	2	1	20
2	2	2	0	25
2	1	1	00	30
1	0	00	0000	40
1	0	00	0000	50
1	00	000	250mcm	60
0	000	0000	300mcm	75

* Main Disconnect is sized for the total output load horsepower on Rotary converters

CAPACITOR CONNECTIONS

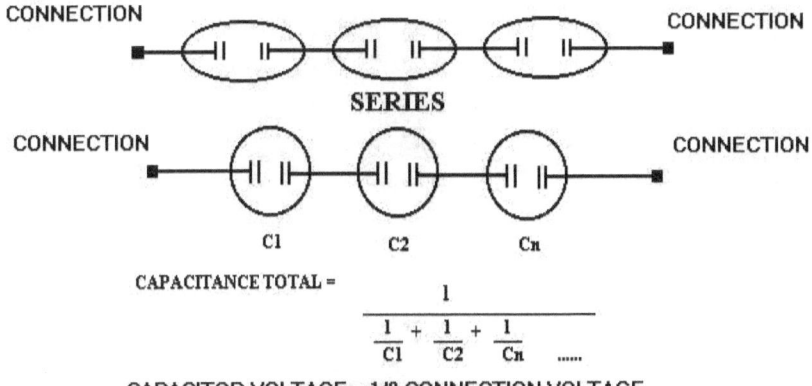

CAPACITANCE TOTAL = $\dfrac{1}{\dfrac{1}{C1} + \dfrac{1}{C2} + \dfrac{1}{Cn} \dots}$

CAPACITOR VOLTAGE = 1/2 CONNECTION VOLTAGE

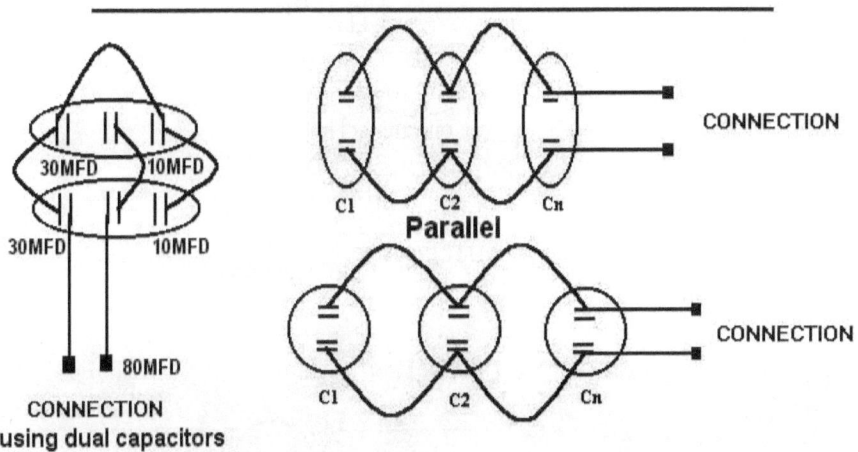

using dual capacitors

CAPACITANCE TOTAL = C1 + C2 + Cn ......

CAPACITOR VOLTAGE = CONNECTION VOLTAGE

METHOD 1 DIAGRAM

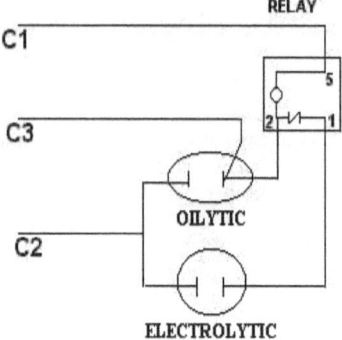

METHOD 2 DIAGRAM

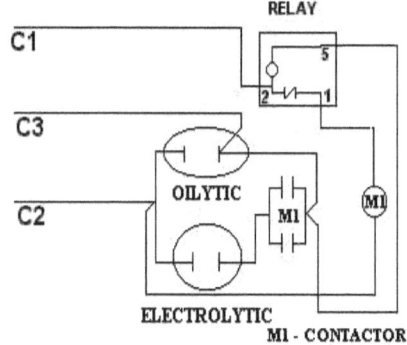

M1 - CONTACTOR

METHOD 3 DIAGRAM

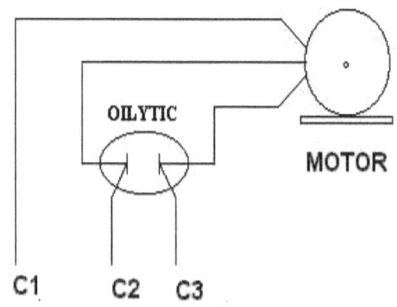

STATIC PHASE CONVERTER CONNECTIONS
SINGLE PHASE ON-OFF OPERATION

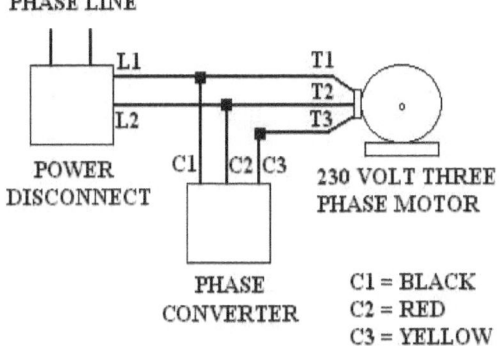

C1 = BLACK
C2 = RED
C3 = YELLOW

* Power Disconnect may also be any two pole switching device sized for the motor load to remove power from motor.

* To reverse rotation of motor, interchange any two motor leads.

* Be sure Converter leads are connected directly to Motor leads or Phase Converter will not operate correctly.

DUAL ROTATION ON-OFF OPERATION

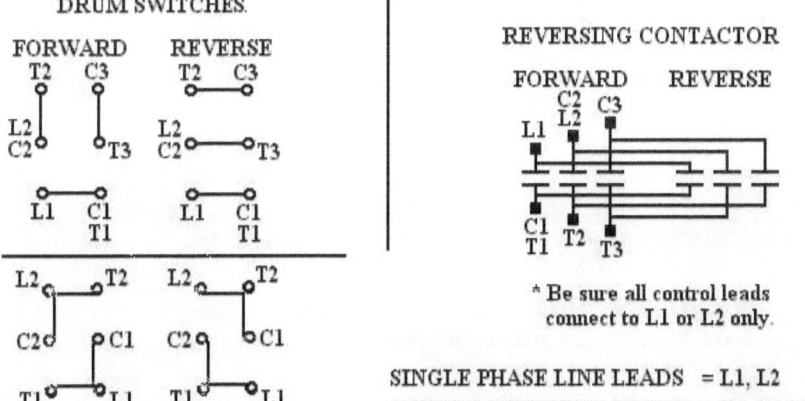

* Be sure all control leads connect to L1 or L2 only.

SINGLE PHASE LINE LEADS = L1, L2
PHASE CONVERTER LEADS = C1, C2, C3
MOTOR LEADS = T1, T2, T3

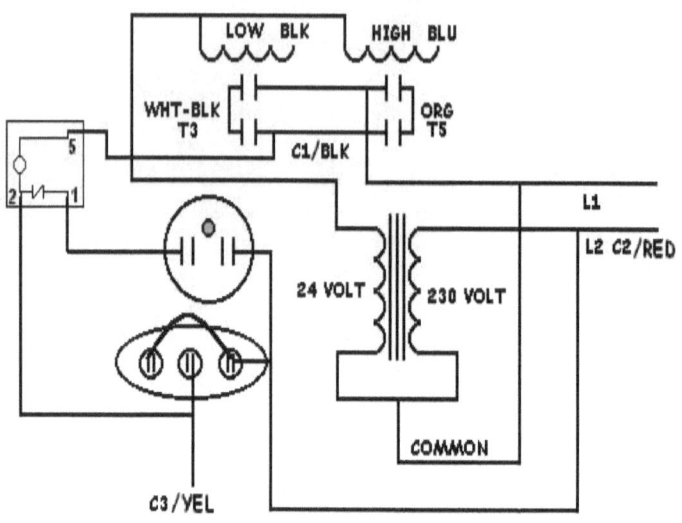

	1-DELTA LOW		HIGH SPEED	2-STAR HIGH
LOW SPEED				
T1-LINE	L1 o	o 5 HIGH BLU	T4-LINE	o——o
T2-LINE	LOW BLK o	o T4	T5-LINE	o o
T3-LINE			T6-LINE	o o
	T2 o —— o L2/C2			
T4-OPEN				
T5-OPEN	T3 o	o T1	T1-T2-T3-CONNECTED	o——o
T6-OPEN	T6 o	o C3-YEL		o——o

Constant Torque 2-SPEED 3600/1800 RPM
PHASE CONVERTER AND SPEED CHANGE RELAYS

ROTARY CONVERTER GENERATOR CONNECTIONS

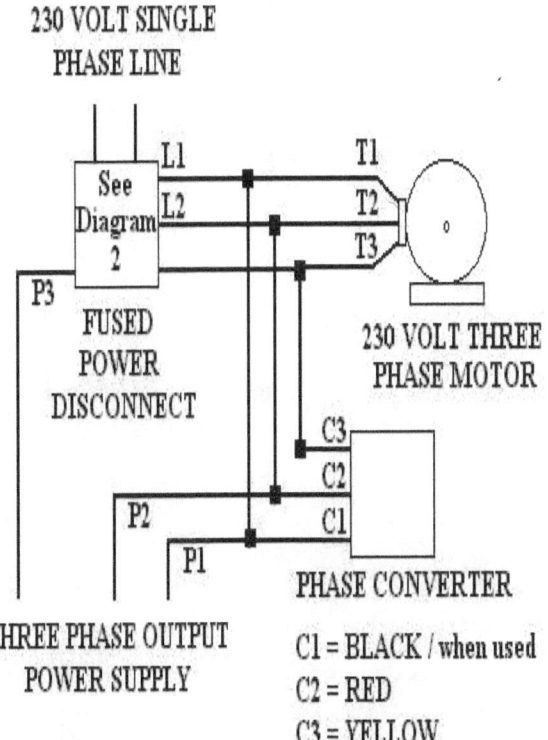

* Some motor designs require a starting capacitance to create a large enough phase shift to allow self starting. These converter boxes have a Black / C1 lead to connect.

* All connections to output three phase power supply should be done according to standard three phase trade wiring practices using fused disconnects and motor starters on each piece of equipment installed.

Rotary Horsepower	Max Starting HP	Combined HP	Output Amps
1 1/2	1 1/2	3	10
2	2	4	14
3	3	6	20
5	5	10	30
7 1/2	7 1/2	15	45
10	10	20	55
15	15	30	85
20	20	40	105
25	25	50	135
30	30	60	160
40	40	80	205
50	50	100	260

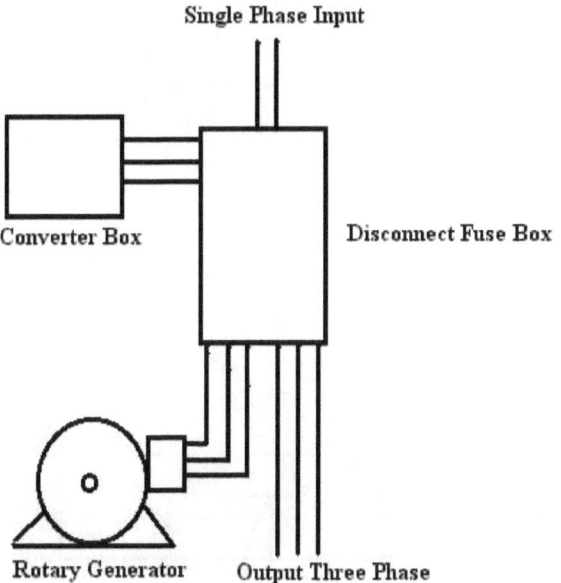

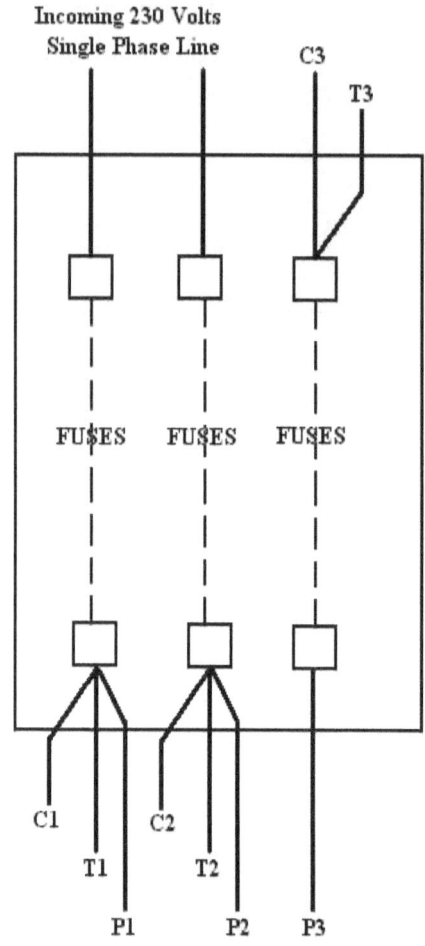

Incoming 230 Volts Single Phase Line

3 - Pole
Fused Disconnect Box
(used as on-off device)

Output Three Phase

Motor Leads T1, T2, T3
Output Three Phase P1, P2, P3

Converter Box Leads C1, C2, C3
C1=Black
C2=Red
C3=Yellow

Typical Connections made at the disconnect panel, note panel size should be rated at the maximum output current and fused accordingly.

Static Connection Diagram

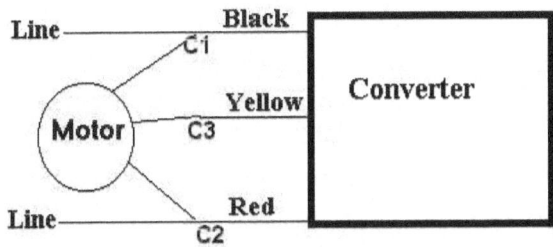

IMPORTANT: DOUBLE CHECK ALL CONNECTIONS

2 Horsepower Plans

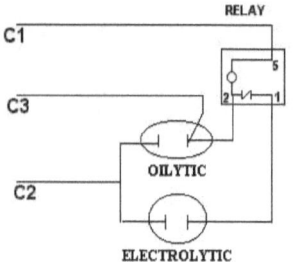

METHOD 1 DIAGRAM

Parts List for 2 HP Rotary Converter

1 - Hotwire Potential Relay
250v - 330v Continuous Voltage
(used on air conditioners)

Starting Capacitance (electrolytic)
100 MFD total
250 VAC - 330 VAC

Running Capacitance (Oilytic)
 70 MFD total
250 VAC - 330 VAC

#14 Hookup wire

Rotary

METHOD 1 DIAGRAM

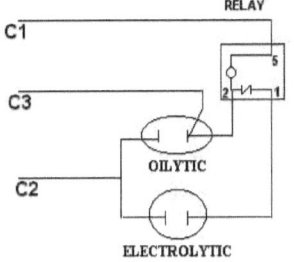

Parts List for 2 HP Static Converter

1 - Hotwire Potential Relay
250v - 330v Continuous Voltage
(used on air conditioners)

Static

Starting Capacitance (electrolytic)
200 MFD total
250 VAC - 330 VAC

Running Capacitance (Oilytic)
70 MFD total
250 VAC - 330 VAC

#14 Hookup wire

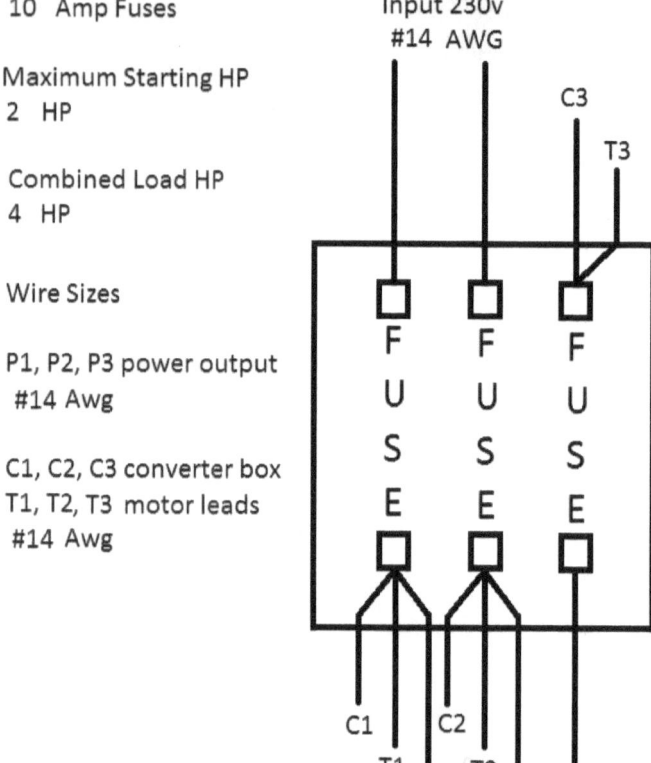

3 Horsepower Plans

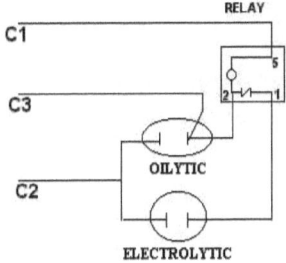

METHOD 1 DIAGRAM

Parts List for 3 HP Rotary Converter

1 - Hotwire Potential Relay
250v - 330v Continuous Voltage
(used on air conditioners)

Starting Capacitance (electrolytic)
150 MFD total
250 VAC - 330 VAC

Running Capacitance (Oilytic)
100 MFD total
250 VAC - 330 VAC

#14 Hookup wire

Rotary

METHOD 1 DIAGRAM

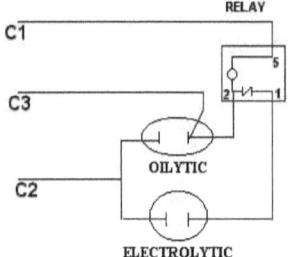

Parts List for 3 HP Static Converter

1 - Hotwire Potential Relay
250v - 330v Continuous Voltage
(used on air conditioners)

Static

Starting Capacitance (electrolytic)
315 MFD total
250 VAC - 330 VAC

Running Capacitance (Oilytic)
100 MFD total
250 VAC - 330 VAC

#14 Hookup wire

Rotary Connection Diagram
3 Pole Fused Disconnect
(used as on-off switch)

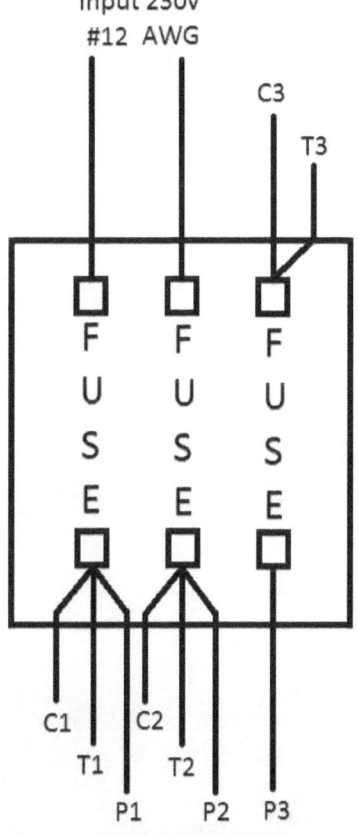

20 Amp Fuses

Maximum Starting HP
3 HP

Combined Load HP
6 HP

Wire Sizes

P1, P2, P3 power output
#12 Awg

C1, C2, C3 converter box
T1, T2, T3 motor leads
#14 Awg

Caution: Use Disconnect rated for maximum load,
Double check all connections

5 Horsepower Plans

METHOD 1 DIAGRAM

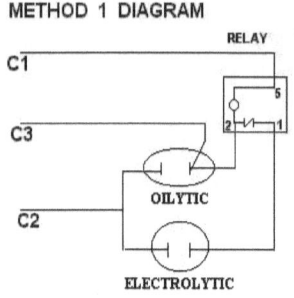

Parts List for   5 HP Rotary Converter

1 - Hotwire Potential Relay
250v - 330v Continuous Voltage
(used on air conditioners)

Starting Capacitance (electrolytic)
250 MFD total
250 VAC - 330 VAC

Running Capacitance (Oilytic)
120 MFD total
250 VAC - 330 VAC

#12 Hookup wire

Rotary

METHOD 1 DIAGRAM

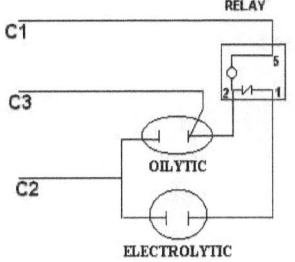

Static

Parts List for 5 HP Static Converter

1 - Hotwire Potential Relay
250v - 330v Continuous Voltage
(used on air conditioners)

Starting Capacitance (electrolytic)
500 MFD total
250 VAC - 330 VAC

Running Capacitance (Oilytic)
150 MFD total
250 VAC - 330 VAC

#12 Hookup wire

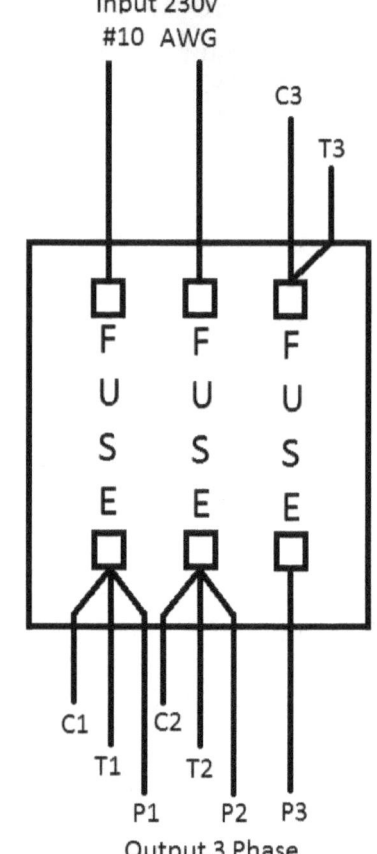

Rotary Connection Diagram
3 Pole Fused Disconnect
(used as on-off switch)

30 Amp Fuses

Maximum Starting HP
5 HP

Combined Load HP
10 HP

Wire Sizes

P1, P2, P3 power output
#10 Awg

C1, C2, C3 converter box
T1, T2, T3 motor leads
#12 Awg

Input 230v
#10 AWG

Output 3 Phase

Caution: Use Disconnect rated for maximum load,
Double check all connections

7.5 Horsepower Plans

METHOD 1 DIAGRAM

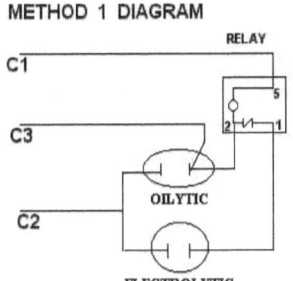

Parts List for 7.5 HP Rotary Converter

1 - Hotwire Potential Relay
250v - 330v Continuous Voltage
(used on air conditioners)

Starting Capacitance (electrolytic)
300 MFD total
250 VAC - 330 VAC

Running Capacitance (Oilytic)
150 MFD total
250 VAC - 330 VAC

#12 Hookup wire

Rotary

Parts List for 7.5 HP Static Converter

1 - Hotwire Potential Relay
250v - 330v Continuous Voltage
(used on air conditioners)

1 - Contactor Relay
2 Pole 30 Amp
230v coil

METHOD 2 DIAGRAM

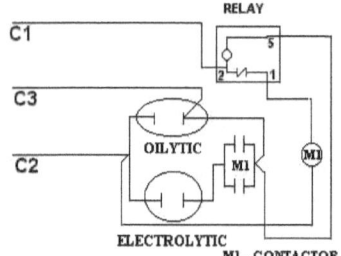

Starting Capacitance (electrolytic)
725 MFD total
250 VAC - 330 VAC

Running Capacitance (Oilytic)
230 MFD total
250 VAC - 330 VAC

#12 Hookup wire

Static

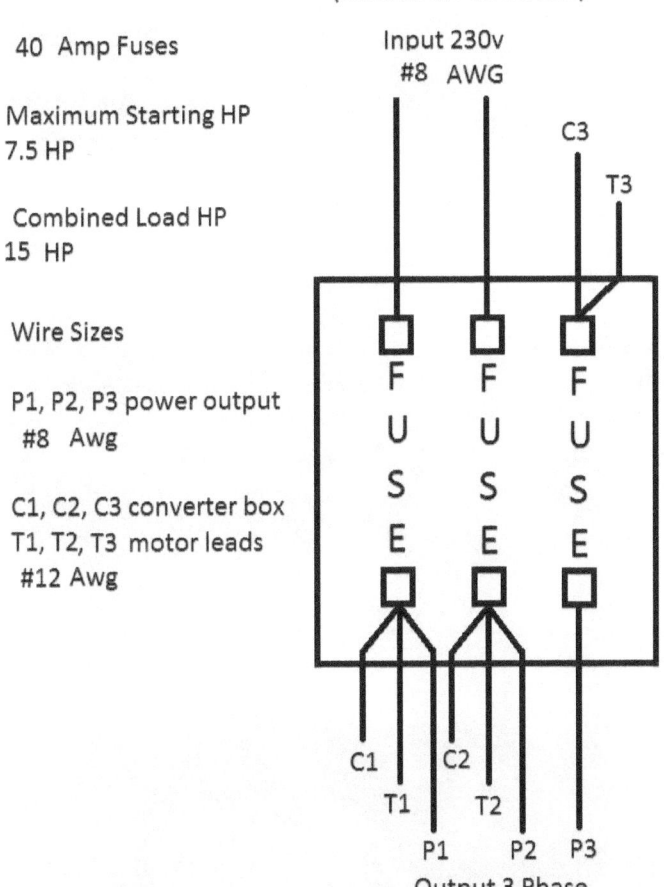

10 Horsepower Plans

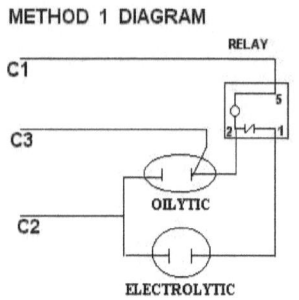

Parts List for 10 HP Rotary Converter

1 - Hotwire Potential Relay
250v - 330v Continuous Voltage
(used on air conditioners)

Starting Capacitance (electrolytic)
450 MFD total
250 VAC - 330 VAC

Running Capacitance (Oilytic)
200 MFD total
250 VAC - 330 VAC

#12 Hookup wire

Rotary

Parts List for 10 HP Static Converter

1 - Hotwire Potential Relay
250v - 330v Continuous Voltage
(used on air conditioners)

1 - Contactor Relay
2 Pole 30 Amp
230v coil

METHOD 2 DIAGRAM

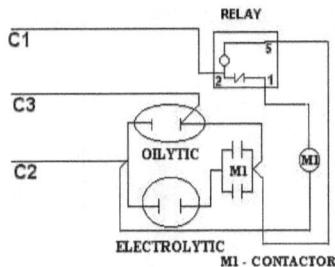

Starting Capacitance (electrolytic)
900 MFD total
250 VAC - 330 VAC

Running Capacitance (Oilytic)
300 MFD total
250 VAC - 330 VAC

#10 Hookup wire

Static

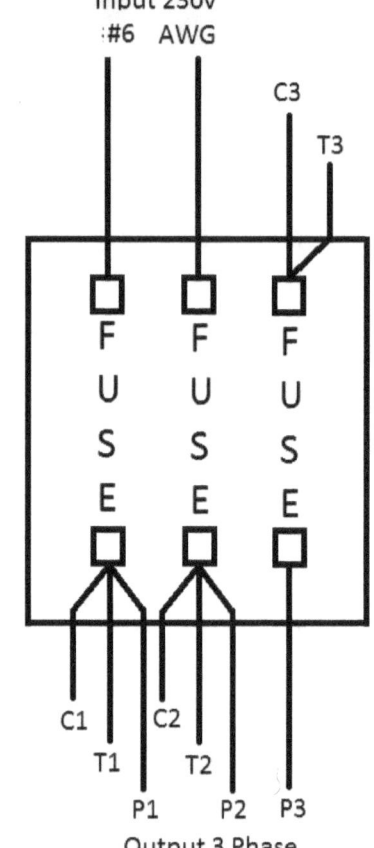

Rotary Connection Diagram
3 Pole Fused Disconnect
(used as on-off switch)

60 Amp Fuses

Maximum Starting HP
10 HP

Combined Load HP
20 HP

Wire Sizes

P1, P2, P3 power output
#6 Awg

C1, C2, C3 converter box
T1, T2, T3 motor leads
#12 Awg

Caution: Use Disconnect rated for maximum load,
Double check all connections

15 Horsepower Plans

METHOD 1 DIAGRAM

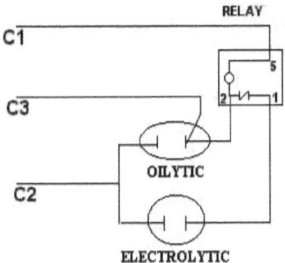

Parts List for 15 HP Rotary Converter

1 - Hotwire Potential Relay
250v - 330v Continuous Voltage
(used on air conditioners)

Starting Capacitance (electrolytic)
600 MFD total
250 VAC - 330 VAC

Running Capacitance (Oilytic)
400 MFD total
250 VAC - 330 VAC

#10 Hookup wire

Rotary

Connect your Running (Oilytic) Capacitors in banks of not more than 200 MFD per bank using #12 AWG wire.

Parallel connections add capacitance values.

Use # 12 AWG wire to connect Starting (Electrolytic) Capacitors. (if used)

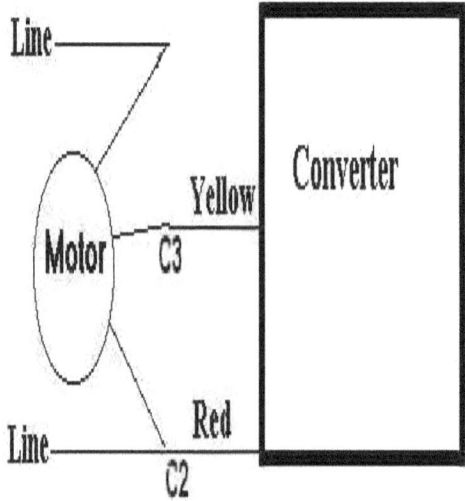

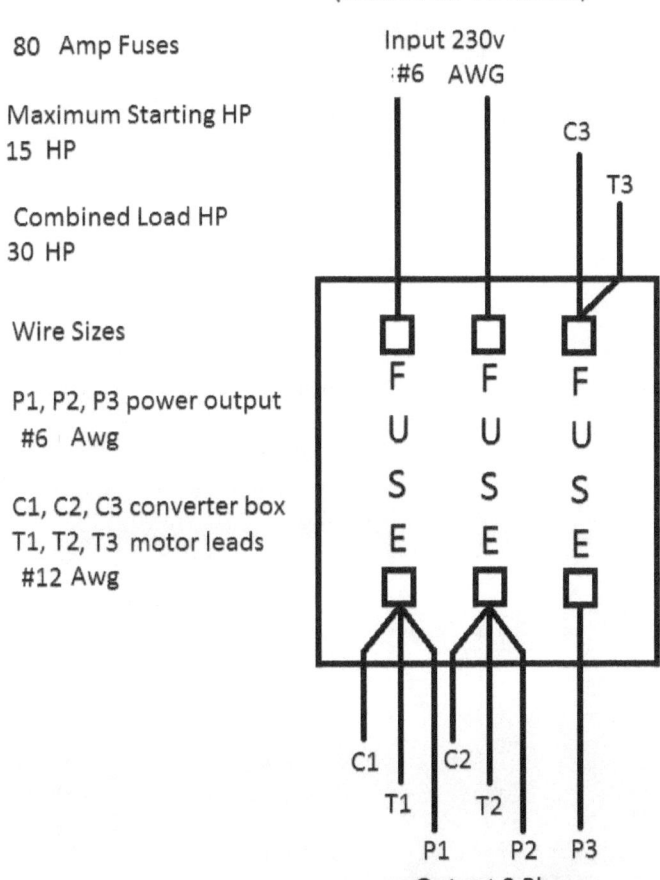

20 Horsepower Plans

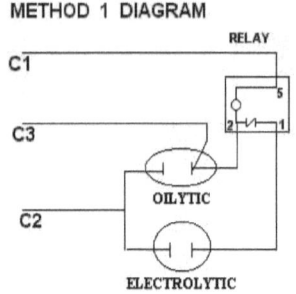

Parts List for 20 HP Rotary Converter

1 - Hotwire Potential Relay
250v - 330v Continuous Voltage
(used on air conditioners)

Starting Capacitance (electrolytic)
600 MFD total
250 VAC - 330 VAC

Running Capacitance (Oilytic)
600 MFD total
250 VAC - 330 VAC

#8 Hookup wire

Rotary

Connect your Running (Oilytic) Capacitors in banks of not more than 200 MFD per bank using #12 AWG wire.
Parallel connections add capacitance values.
Use # 12 AWG wire to connect Starting (Electrolytic) Capacitors. (if used)

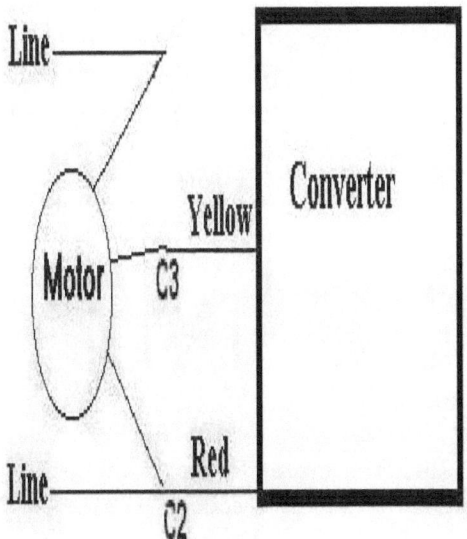

Rotary Connection Diagram
3 Pole Fused Disconnect
(used as on-off switch)

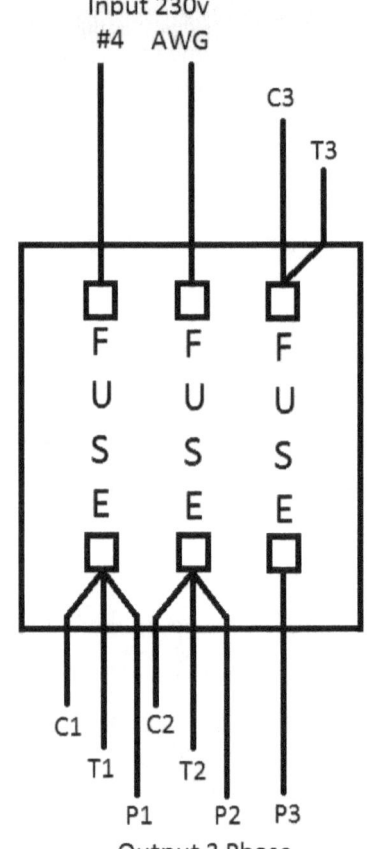

100 Amp Fuses

Maximum Starting HP
20 HP

Combined Load HP
40 HP

Wire Sizes

P1, P2, P3 power output
#4 Awg

C1
#12 Awg

C2, C3 converter box
#8 Awg

T1, T2, T3 motor leads
#6 Awg

Caution: Use Disconnect rated for maximum load, Double check all connections

25 Horsepower Plans

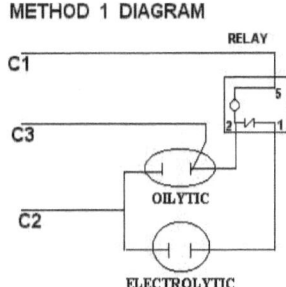

METHOD 1 DIAGRAM

Parts List for 25 HP Rotary Converter

1 - Hotwire Potential Relay
250v - 330v Continuous Voltage
(used on air conditioners)

Starting Capacitance (electrolytic)
600 MFD total
250 VAC - 330 VAC

Running Capacitance (Oilytic)
750 MFD total
250 VAC - 330 VAC

#6   Hookup wire

Rotary

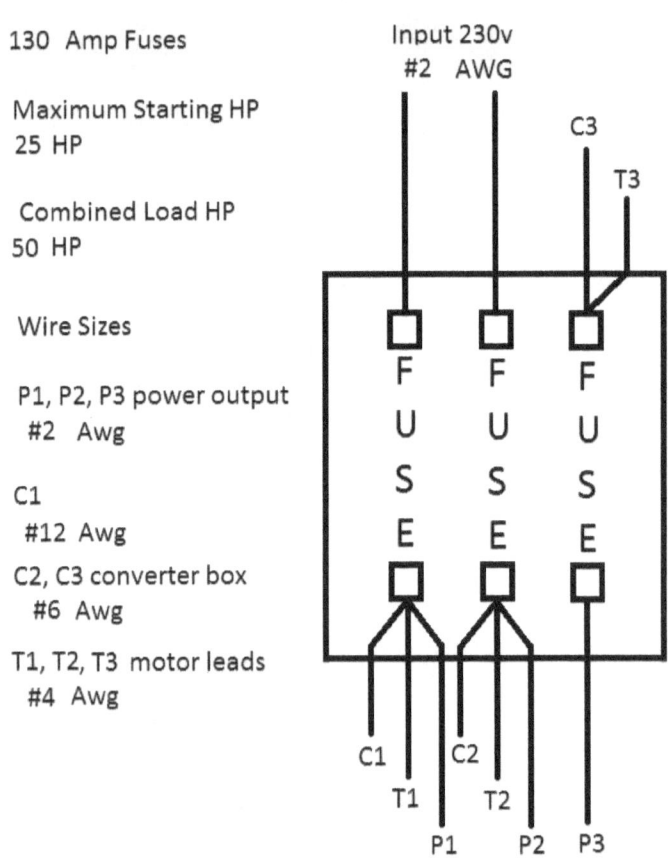

30 Horsepower Plans

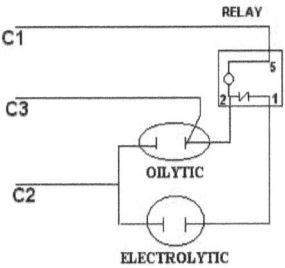

METHOD 1 DIAGRAM

Parts List for 30 HP Rotary Converter

1 - Hotwire Potential Relay
250v - 330v Continuous Voltage
(used on air conditioners)

Starting Capacitance (electrolytic)
600 MFD total
250 VAC - 330 VAC

Running Capacitance (Oilytic)
900 MFD total
250 VAC - 330 VAC

#4   Hookup wire

Rotary

Rotary Connection Diagram
3 Pole Fused Disconnect
(used as on-off switch)

150 Amp Fuses

Maximum Starting HP
30 HP

Combined Load HP
60 HP

Wire Sizes

P1, P2, P3 power output
 #1 Awg

C1
 #12 Awg

C2, C3 converter box
 #4 Awg

T1, T2, T3 motor leads
 #2 Awg

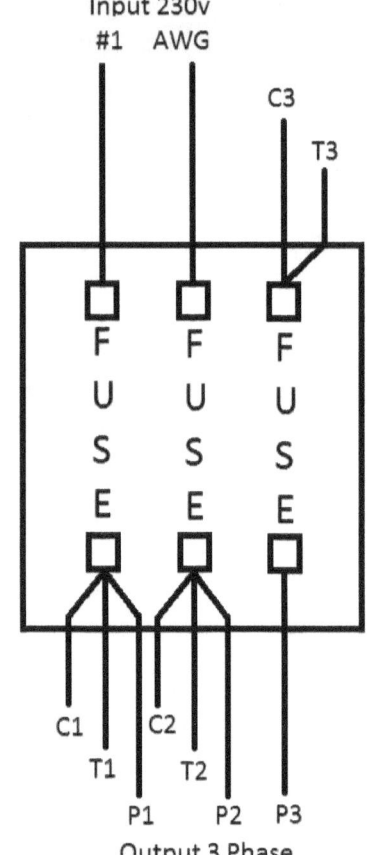

Caution: Use Disconnect rated for maximum load,
Double check all connections

Connect your Running (Oilytic) Capacitors in banks of not more than 200 MFD per bank using #12 AWG wire.
Parallel connections add capacitance values.
Use # 12 AWG wire to connect Starting (Electrolytic) Capacitors. (if used)

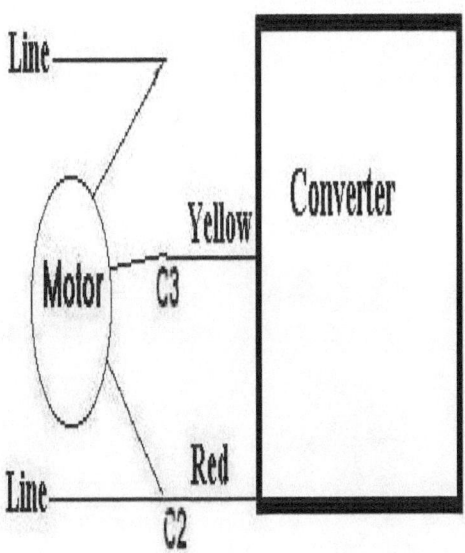